Wired to the World,
Chained to the Home

Penny Gurstein

Wired to the World, Chained to the Home: Telework in Daily Life

UBCPress · Vancouver · Toronto

Printed in Canada on acid-free paper ∞

ISBN 0-7748-0846-2

National Library of Canada Cataloguing in Publication Data

Gurstein, Penelope Cheryl
  Wired to the world, chained to the home

  Includes bibliographical references and index.
  ISBN 0-7748-0846-2

1. Telecommuting. 2. Home labor. 3. Home-based businesses. I. Title.
HD2336.3.G87 2001 331.25   C2001-910230-5

This book has been published with the help of a grant from the Humanities and Social Sciences Federation of Canada, using funds provided by the Social Sciences and Humanities Research Council of Canada.

UBC Press acknowledges the financial support of the Government of Canada through the Book Publishing Industry Development Program (BPIDP) for our publishing activities.

Canadä

We also gratefully acknowledge the support of the Canada Council for the Arts for our publishing program, as well as the support of the British Columbia Arts Council.

Printed and bound in Canada by Friesens
Set in Stone by Artegraphica Design Co.
Copy editor: Sarah Wight
Proofreader: Tara Tovell
Indexer: Patricia Buchanan

UBC Press
The University of British Columbia
2029 West Mall
Vancouver, BC V6T 1Z2
(604) 822-5959
Fax: (604) 822-6083
E-mail: info@ubcpress.ubc.ca
www.ubcpress.ca

Contents

Figures and Tables

Tables

Acknowledgments

Undertaking as large a project as a book can tax all of your resources and is virtually impossible without the support and guidance of others. I would like to thank the funding sources that allowed me to carry out this research. Canada Mortgage and Housing Corporation (CMHC), a Crown corporation of the federal government of Canada, funded a doctoral scholarship for the research for the 1990 study conducted in northern California. The 1995 survey by the author, "Planning for Telework and Home-Based Employment: A Canadian Survey on Integrating Work into Residential Environments," was also supported by Canada Mortgage and Housing Corporation. I was awarded University of British Columbia Humanities and Social Sciences (HSS) grants in 1993-4 and 1999-2000, funded by the Social Sciences and Humanities Research Council of Canada (SSHRC), which allowed me to update my data and do the Vancouver case study. I would also like to acknowledge the publishing subsidy awarded by the Humanities and Social Sciences Federation of Canada Aid to Scholarly Publications Program for this manuscript.

I would like to acknowledge Professor Manuel Castells, who was highly supportive of my research on telework in the early stages of its development at the University of California, Berkeley, and whose own work has continued to inspire me. I would also like to acknowledge the anonymous reviewers of this manuscript who helped me focus my thoughts. I would especially like to thank my editor at UBC Press, Emily Andrew, who helped me conceptualize the framework for this research, gently prodded me into completing the manuscript, and expertly shepherded it through the publishing process.

Numerous graduate research students at the School of Community and Regional Planning at the University of British Columbia worked on the various studies that comprise this book, and I want to acknowledge their contributions. For the 1995 Canada-wide study on telework, David Marlor, Shauna Brail, Janice Keyes, and Zarina Mulla assisted me in the data input and analysis. For the 2000 Vancouver case study, Theresa Harding assisted in organizing the study and Andrew Ramlo graphically presented the greater Vancouver area statistics on home-based work. Leanne Sexsmith and Paul Faibish assisted in the literature review. Deana Grinnell assisted in the final stages of completing the manuscript.

I want to also acknowledge the continuing support of my family and friends, especially my parents, Emanuel and Sylvia Gurstein, and the patience of my daughter, Natasha.

Finally, I would like to acknowledge the people I interviewed, who welcomed me into their homes and shared their thoughts. This is their story that I am telling and without their contributions this book wouldn't have been possible. In order to protect the privacy of the study respondents all names have been changed.

Wired to the World,
Chained to the Home

1
Telework As Restructured Work

> The issues of technological restructuring are altering not
> just our jobs and our work, but our language, consciousness,
> and identity (Menzies 1996, xiv).

In the ten years that I have been investigating telework I have grown increasingly sceptical of the message that promotes technology as the panacea for the drudgeries of work. As technology, in the form of computers and advanced information and telecommunications systems, permeates every aspect of society, there seems to be little critical discourse on its impact on the way we conduct our daily lives. A television commercial for a new computer a few years back typifies the intense mythologizing that surrounds our use of technology: a woman, professionally dressed in a business suit, is seated in front of her computer concentrated on work while several apparently happy children play at her feet. Fast forward to the year 2000 and a photograph accompanying a newspaper article offers a similar message: a contented home-based worker sits in front of her computer surrounded by her playing children (Gram 2000). The implicit messages both of these images conveyed to me were first, "We (women) can have it all," and second, "Technology is a benign tool." While my own experience has demonstrated to me the difficulty of combining work and family life and the powerful influence that technology has in ordering spatial, temporal, work, and interpersonal relations, few studies have investigated this on a household level.

In the restructuring of work that is occurring to address the socioeconomic priorities of the economy and workforce, telework has become a prominent strategy for employers and policy makers. Telework, typified as work performed with the help of information and communication technologies, often located at a distance from a main office site, includes a range of working relationships: employees connected to corporate networks while working from their homes or other remote locations, such as telecentres or client offices; self-employed consultants usually working from home, or home-based business operators operating businesses from their homes; independent contractors or self-employed subcontractors who rely on ICTs (information and communication technologies) in order to carry out their work; and workers, whether directly employed or outsourced, located in back offices or call centres, linked telematically to employers' central offices. The types of work and the locations that permit telework are highly conducive to meeting the increased demand for flexible work arrangements by both workers and corporations.

Telework (or "telecommuting" as it also called in the United States), as distinct from other forms of work based in the home, is defined as work-related substitutions of telecommunications and related information technologies for travel (Huws, Korte, and Robinson 1990). Telecommuting came into prominence in the 1970s as a work option that reduces dependence on transportation (Mokhtarian 1991a; Nilles et al. 1976), but it is of interest now to both the private and public sectors because it produces a mobile, flexible labour force and reduces overhead costs (Huws 1991). Neither of these terms always implies working at home, as satellite offices or neighbourhood telework centres close to employees' homes, equipped with telecommunications equipment and services, can substitute for the commute to a centralized office.

Any large-scale telework movement can be attributed to existing economic conditions as well as technological advances. The internationalization of the economy has forced North American companies to try to cut labour costs to compete against companies who are producing cheaper, better quality goods elsewhere. The transformation from an industrial to a service economy has created many jobs that can be done independently of a centralized facility. Technological change in the form of advanced telecommunications technologies has made it possible to transport a variety of jobs, including data entry, offshore. Offices in North America are increasingly becoming automated, with a projected loss of management and clerical jobs. In order to remain competitive, companies are creating a two-tiered workforce of core and peripheral workers. While a core of full-time salaried workers remains, temporary workers are hired on a contingency basis. For many of these workers, the home becomes their work site.

The family is also undergoing structural changes that are contributing to the increase in telework. With dual-earner or female-headed families increasingly becoming the norm, the boundaries between work and family have changed. The burden of unpaid domestic labour, however, still falls primarily on women. Because of their double burden, women have sought flexible ways to work, including part-time work and self-employment. Currently, woman-owned businesses are the fastest growing segment of the small business population. The home provides the workplace for many of these women in business for themselves.

Two critical factors in the analysis of telework are who controls the information (i.e., who sends information to whom) and how the technologies can be manipulated. The activities generated by information technologies are part of larger societal processes that are locally situated in the home.

There are now numerous modes of electronic communication between individuals, corporations, and the global society that can originate from the home or other mobile work sites. Some of these interactions allow for decentralized information gathering and decision making, while others reinforce centralized and hierarchical structures. Activity patterns depend upon an individual's economic and information resources, and on the temporal and spatial constraints imposed on him or her.

This book analyzes the experiences and practices of teleworkers, including employees, independent contractors, and self-employed entrepreneurs, who use ICTs in the course of their work. It describes the socioeconomic environment of "flexible" employment and economic restructuring under which this form of work is being generated in North America. Recognizing that diverse forms of microentrepreneurship and home-based work are well-established economic strategies, I strive here to articulate the differences and similarities between informational at-home and/or distance work and other forms of work.

I argue against a technological determinist stance that obscures a class and gender analysis, placing telework within the framework of the different work relationships that affect an individual's economic and information resources and the temporal and spatial constraints imposed by household responsibilities. Gender becomes significant in understanding the experiences of teleworkers because gender differences are prevalent in the employment status of homeworkers. Moreover, often female at-home workers have the dual responsibilities of paid work and family, while male home-based workers primarily view themselves as engaged only in paid work. These different sets of perceived responsibilities affect the experience of working at home.

The discussion in the rest of this chapter locates telework within the context of the flexible labour market precipitated by changing socioeconomic and domestic priorities. It is shown that women are the most affected by this labour market and the most vulnerable to its consequences. In addition, women's daily life patterns are influenced by the private/public dualism manifested in our cities, which impedes the legitimization of women's home-based work activities. In reconceiving these dualities I argue that these distinctions do not represent people's lived experiences, and that the roles assigned based on these dualities do not reflect the fluidity of our society. Roles may vary situationally, but it is identities, those attributes of a person that give meaning, that help sustain us in our increasingly complex sets of relationships, especially in work practices. Telework is a particularly interesting example of the immersive nature of the relationship between work and technology in creating "identity." In certain instances, however, telework is also a strategy of resistance to the totalizing nature of work. The assumptions and myths about telework and its causes clearly need to be dissected.

Flexible Labour

Labour flexibility has become a significant trend in workforce profiles in both developed and less developed countries. This flexibility, however, differs under varying labour regimes and at varying skill levels. It is assumed that post-Fordist capitalism requires "flexible specialization" to meet diversified and specialized markets. Burawoy and Lukács argue to the contrary that "flexible specialization under capitalism is less an economic imperative and more a political stratagem to elicit consent in a period when middle management is under assault. It becomes a means of further expropriating control from the direct producer" (1992, 20). Haraway (1990) identifies this situation as the "homework economy." While she includes the literal interpretation of homework, such as home-based electronics assembly, she also broadly defines the restructuring of work with characteristics formerly ascribed only to "female" jobs, such as vulnerability to exploitation and erratic work schedules.

International studies have noted that women generally have been affected differently than men by the fluid economic and social landscape of the restructuring that has been occurring worldwide (Aslanbeigui, Pressman, and Summerfield 1994). Women have had to bear a disproportionate amount of both paid work and unpaid domestic work to maintain their households. Castells (1996) argues that it is their flexibility as workers that has resulted in the expansion of women's paid employment. This flexibility in schedules, and in entry and exit from the labour market, has resulted in women constituting the bulk of part-time and temporary employees, and a significant share of the self-employed. Nevertheless, this flexibility is at a cost in stretched time and resources (Hochschild 1989, 1997).

The gender and ethnicity dimensions of flexible production techniques are reflected in the participation of women and immigrants in the workforce, mostly in lower-paid work involving inferior working conditions (Mitter 1992; Cohen 1991). These gender- and ethnic-specific labour practices often rest on a revival of such techniques as subcontracting and family labour systems that involve patriarchal management structures and homework. Such practices make it easier to substitute lower-paid casual female labour for that of more highly paid and more difficult to lay off core male workers (Harvey 1989). These unregulated jobs are often concentrated in isolated and hidden work sites in homes and garages. The consequence of their invisibility is the prevalence of substandard working conditions and the potential for exploitation from employers and contractors.

Not all flexible workers, however, are as vulnerable as those described above, and homeworkers cannot be treated as a single group. Those teleworkers who are highly skilled and employed, rather than on contract to their employer, appreciate the control over their time that working at home affords them. Carnoy, Castells, and Benner (1997) take a nuanced approach to describe flexible labour in the North American context. This encompasses both voluntary and involuntary labour practices, and includes temporary and part-time workers, the self-employed, and workers employed in business services. While recognizing that often flexible labour is positive for both employers and workers, they point out that it can become onerous for workers "denied access to the standard, or traditional, labour contract" (31). Using a case study of workers in California's Silicon Valley, they show that though flexible employment can benefit mobile, high-skilled workers in ICT professions, the flexibility of low-skilled workers (such as janitors working on contract) benefits companies, resulting in job losses and reductions in wages and working conditions.

A study of low-skilled information-processing workers in less developed countries (Pearson and Mitter 1993) found a predominance of women in low-skilled employment, while the majority of high-skilled workers were men. While conceding that information processing offers the developing world significant employment opportunities, Pearson and Mitter are concerned that "in the long term some jobs will be upgraded and new technical skills and skills combinations will be needed; but other jobs will be downgraded and the conditions of work for 'peripheral' workers are likely to deteriorate" (52). Menzies eloquently describes a similar Canadian telework model of "computer-defined and computer-controlled tasks dispatched to remote worksites" (1996, 110). In her critique of the foundations of this model she states, "They represent a shift in the distribution of power in our society toward computer systems and those who control them, and a new version of class polarization – here across the digital divide of technological enfranchisement or disenfranchisement, of working with computers or working for them. They also represent a new form of social control: from a human context of industrial relations to an almost entirely cybernetic context."

As illustrated by these studies, a consequence of the shrinking of the welfare state and the spread of flexible production techniques has been growing social polarization. This is largely based on the decline in the bargaining power of labour relative to capital and the subsequent bifurcation of the labour force into a small number of "good jobs" (i.e., secure, long-term, well-paying, unionized jobs) and a much larger number of "poor jobs" (i.e., part-time, part-year, low-paying jobs without benefits and unionization) (Duffy, Pupay, and Glenday 1997). Workers in "poor jobs," in the flexible or floating labour force of homeworkers, domestic workers, independent contractors, and agricultural workers, are a sizable percentage of workers worldwide and are rapidly increasing in the North.

The New Frontier: The Mobile Workplace

An estimated one-quarter of the working population in North America does some or all of its paid labour from home or close to home. Many more dream of doing so. Working at home is increasingly mythologized as the new frontier – an individual's ultimate expression of autonomy, freedom, and control – made possible by telecommunications and information technologies. While there is a long and global tradition of home-based workers, the use of computers, modems, and faxes to work at home or close to home, variously known as "the electronic cottage," "electronic homework," "telework," "telecommuting," "networking," "distance work," and "flexiplace," takes on such significance in predictions about the future of work that it is difficult to escape the suspicion that this trend has acquired a symbolic stature beyond its actual prevalence. Excessively optimistic predictions of home-based work do not reveal, however, the millions compelled to work at home by socioeconomic necessity and technological redundancy. For them, home-based work is a survival strategy and a form of resistance to societal forces beyond their control. Bringing work home affects every aspect of their daily lives, blurring boundaries between work and home life, workplace and home, public and private space, and male and female roles.

There is a tremendous amount of hyperbole about the promise of the "information highway" and its impact on daily life and work patterns. Telework and home-based employment offer millions of people liberation from unwanted commutes, more flexibility and control over time and resources, and the fostering of more cohesive communities. This promise, though, must be tempered by the reality of the day-to-day lives of people working in homes and communities often irreconcilable with this vision of the future. The flexible, isolated work site or the mobile workplace has implications for the future of work and societal relations. Telework and home-based employment are work practices related to changes in technology and family life, precipitated by the current global economic restructuring, which has local consequences for the reshaping of spatial forms and social dynamics. The rising number of information-based and service-related occupations and jobs, increasing contract work and part-time employment, widespread use of computers and telecommunications, corporate restructuring, and workers' desire to balance family and work are all factors reinforcing an increase in flexible work patterns.

While existing literature covers a broad spectrum of issues related to telework and home-based employment, from the sweeping speculations of the futurists to dramatic accounts in the popular media, relatively little research deals directly with the issues addressed in this book. This book is about the experiences of telework and home-based employment, and how these forms of work are manifested in people's daily lives and environments. It addresses two questions that pertain to the social and spatial impact of this reconceptualization of work practices: how do people accommodate telework and home-based employment in their use of time and space, and what is the role of home, work, and community life in this context?

It is estimated that in the United States from 20 million to 38 million people, or 30 percent of the US labour force, work at home at least part of the time (Deming 1994; US Department of Transportation 1993). Telework has grown in Canada from 600,000 people in 1991 to one million in 1996 (Statistics Canada 1991a, 1997), and some estimates based on a more all-encompassing definition find two million Canadians – nearly one-quarter of the working population – doing some or all of their paid labour at home (Orser and Foster 1992). While this shift in employment patterns is predicted to have a considerable influence on patterns of daily life, the planning and design of residential communities have not for the most part recognized this occurrence.

Twentieth-century cities have been planned for home/work separation, and in particular planned to make work-related activities efficient based on their separate location. However, residential planning based on the principle of separating home and work activities may be outdated for an increasing number of North American households. When the location of work shifts for a sizable percentage of the labour force, planners and designers need to reconsider the policy and design implications of the home/work relationship. The public and private sectors' increased interest in the nature and extent of paid work conducted at home reflects its potential impact on economic development. Introducing opportunities for work into residential communities in the form of telework/telecommuting, home-based employment, satellite offices, and neighbourhood telework centres has important implications for land use, urban form, the housing industry, transportation, and services. Challenging the live/work dichotomy and reshaping existing development patterns brings the role of communities and the function of regulatory mechanisms under scrutiny. Equally important are the social implications of a work life that is dissociated from an organized workplace and dispersed to a variety of locations.

The home is becoming the nexus for a whole range of activities, making for an increasingly home-based society. A retreat to the home is occurring in the areas of work, socializing, entertainment, and education. This retreat is fuelled by fear and uncertainty about the outside world and by the convenience of technological fixes. Home-based activities that transform the home into a sphere for both production and consumption have the potential to decentralize resources and provide flexibility and control over both work and home life. At the same time, such activities could atomize and isolate homeworkers from interactions in the larger society. The societal consequences of a solitary work life need careful consideration.

While home-based work does have negative effects, at the same time it offers important opportunities to reorganize our homes and communities. Integrating opportunities for work such as telework centres into residential neighbourhoods is a way of revitalizing single-use areas and reducing the enormous energy and transportation inputs required to maintain North American lifestyles. While replanning residential communities with work in mind will not alleviate all of the problems of our increasingly complex environment, it will go far toward creating more sustainable, humane communities.

This book is based on a framework that acknowledges the tensions between individuals as agents of change and the wider social systems and structures in which they are embedded. I argue that though home-based work has become a significant work trend (as recent North American census figures indicate) it is not the panacea for the often conflicting demands of home and work life. However, the importance of exploring the impact of home-based work on urban life should not be underestimated. Home-based work challenges existing conceptualizations of work and domestic life, and public and private space, leading to new interpretations of the urban realm. To this end, this book contributes to the debate on the rethinking of our socioeconomic and environmental priorities in the urban sphere by investigating the impact of telework and home-based employment on daily life patterns and the use of homes and neighbourhoods, and examining the implications for the planning and design of homes and communities.

The Private/Public Dualism

This book is situated within the growing literature on gender-sensitive planning and design. For the past thirty years, feminist planners have provided a valuable critique of traditional planning practice, rethinking what knowledge is relevant and whose actions we are planning for (Sandercock 1998; Eichler 1995; Greed 1994; Sandercock and Forsyth 1992; Moore Milroy 1991). By emphasizing a multiplicity of experiences, they have identified the importance of reorienting community planning practice from a focus on land use to a more holistic approach that includes issues such as child care, mass transit, safety, and affordable housing. Feminist theoreticians have, for example, drawn attention to the male bias in traditional land use planning (Hayden 1981, 1984; Mackenzie 1988). They challenge modern North American cities excessively codified by zoning restrictions, and dualistic divisions between private and public space, and between home and work environments (Ritzdorf 1990). The benefits of such gender-sensitive approaches to planning are meant to accrue to the entire population, not just women.

With the redefinition of women's and men's roles, as is presently occurring with the dual-career household and multiple definitions of family, approaches to planning practice have to be cognizant of the variety of roles and of the fact that these roles are socially constructed and constantly changing. By considering the needs and preferences of diverse groups, planning can more significantly address their concerns. Correspondingly, the separate sphere ideology that defines the home in opposition to work and as a refuge from the public world is being reexamined with the advent of technological, economic, and social restructuring. Cities are now being characterized in terms of the interconnections between public and private spheres and between socioeconomic and cultural phenomena (Andrew 1992; Moore Milroy and Wismer 1994).

Such approaches acknowledge biases in decision making; planning is not value-free but very much ideologically bound (Little 1994). This acknowledgment is an important step in breaking down the belief that planning is objectively neutral. These approaches also recognize the diversity of human experience and the variety of ways of knowing and constructing knowledge. In this context, communities are not just spatially defined but are sets of resources that provide opportunities and impose constraints. The feminist approach to architecture defined by Franck (1989) includes many elements desirable in planning as well: connectedness and inclusiveness, an ethic of care and value for everyday life, values of subjectivity and feelings, values of complexity and flexibility, and cooperation and collaboration. While many of these concepts may appear to be utopian, when examined closely they constitute a way of conceiving social and spatial relations that challenge the socially imposed dualities within the urban realm.

Moore Milroy (1991) maintains that planning decisions in North America are based on a narrow definition of work that elevates the waged form of work, done at particular times of the day and week and in specific locations, to a higher status. In this perspective, cities are planned as places of work, and neighbourhoods as residual places. Limiting the conceptualization of work to formal employment ignores the unpaid work done in homes and communities, and the increasing diversity of paid work done at home and in mobile locations. This bias defining work as separate from domestic activities is a fundamental organizing element of urban structures, codified by zoning restrictions. Pateman (1989) argues, as many do, that the public and private spheres are inextricably connected and interdependent. Their interdependence, however, is not mutually complementary but hierarchical, based on a relationship where the public sphere exerts more power in decisions and actions (Moore Milroy and Wismer 1994).

Boris (1994) argues that such dualities as home and work, and private and public, are false dichotomies that impede recognition of home-based working women as wage earners. Family obligations, power relations within a family, class, and race, as well as public policies such as taxation, immigration, and welfare, affect a woman's position in the labour market. She further concludes, "By sustaining the association of women with the home, policymakers place a barrier in the way of reorganizing social life to recognize the earning and nurturing obligations of us all" (365). An examination of the movement toward telework by individuals, private corporations, and public institutions illustrates the interconnections between the private and public spheres, and the hierarchical relationship between them.

Identity, Power, and Resistance

Haraway dissects these dualisms further. She argues that the distinction between public and private domains in characterizing women's lives is totally misleading and instead suggests a network or integrated circuit image that encompasses spaces, identities, and "the permeability of boundaries in the personal body and in the body politic" (Haraway 1990, 212). Her "cyborg" image, a hybrid of machine and organism, points the way to a new theoretical understanding of feminism that embraces technology in order to challenge the "informatics of domination" (223). Butler (1993) challenges the notion of binarism even more strongly, arguing that such a notion operates counterproductively in gender and feminist discourse and asserting that individuals have the power to refuse to play their gender roles, thus subverting power systems. She claims that identities do not preexist their "performance," thus emphasizing human agency and the importance of individual actions and resistance. Butler stresses the lack of a stable self and the fluidity of power.

Castells, while concurring that identity is important in understanding contemporary society, offers a different interpretation. Identities, Castells asserts, are distinct from roles in that roles are "defined by norms structured by the institutions and organizations of society" while identities are created by "the construction of meaning on the basis of a cultural attribute, or related set of cultural attributes, that is/are given priority over other sources of meaning" (Castells 1997, 7, 6). Roles vary situationally, but in a globally fluid networked society identities are sustained over the time-space spectrum. Castells typifies identity-building in three distinct ways: "legitimizing identity," "resistance identity," and "project identity." "Legitimizing identity" is a strategy of the dominant institutions of society to extend their domination, "resistance identity" seeks to subvert the dominant institutions, and "project identity" seeks to transform them.

Butler's and Castells' different arguments both critique Goffman's (1959) analysis of roles, most importantly in their understanding of resistance. Goffman argues in his discussion of roles and role settings in daily work encounters that work is a performance with an explicit script governing behaviour. In such an environment an alienated worker can potentially be liberated through an improved work culture and workspace. However, our greater understanding of human agency in the making of meaning diverges from this interpretation (Giddens 1991). Structuration theory, focusing on the relationship between human agents and the wider social systems and structures in which they are embedded, objects to the portrayal of people as passive actors within determined structures. Instead, it recognizes the importance of the knowledgeable human subject as an agent of change while at the same time examining the structures within which the human agent operates. It is through the specific acts of individuals that meanings are created in society.

Resistance strategies are part of that meaning making. Resistance, as tactics focused on the self-construction of a new identity rooted in human dignity, is a continuing theme in discussions of work practices (Scott 1990; Ong 1987). The Luddites, a nineteenth-century society of textile workers who tried to sabotage the mechanization of their industry by breaking their machines, are still recalled when people question the increasing use of technology as a substitute for human labour (Bailey 1998). Workers in situations in which they have very little control can manifest resistance through a whole host of behaviours. For example, Koch-Schulte's study (2000) of resistance among call centre workers found that they used worker-managed computer pacing when the stress became too onerous.

Power is central to any discussion of resistant work practices. Foucault (1980) argues that power is diffuse. It is the product of a "net-like organization" that is "something which circulates, or rather as something which only functions in the form of a chain ... In other words, individuals are the vehicles of power, not its point of application" (98). Zuboff (1988) applies this analysis to workplaces relying on computer-based technologies and finds that "authority is used to shape conduct and sensibility in ways that contribute to the maintenance of current configurations of power" (1988, 222). Industrial factories were places where bodily discipline, regulation, and surveillance were taken for granted, but the monitoring ability of information technologies renders the physical presence of authority unnecessary. Increasingly we have to ask the question, Are we working *with* computers or working *for* them?

Our identities, as formed by the imperatives of a global economy, are now immersed in our relationship to our technologies. Telework and other forms of flexible labour fit into these new conceptualizations of workers. Nevertheless, within the apparently all-encompassing dominance of the global economy, individuals are exhibiting agency in navigating their lives and controlling how and where work is to be conducted. However, although resistance is integral to the social practices of teleworkers, any fundamental change will depend on transformations in the societal meanings and practices of work.

Assumptions about Telework

Among the assumptions about the trend toward telework that need to be analyzed is the assumption that the computer determines the decision to work at home. While the computer is enabling millions of people to work at home, the reasons that they are at home stem more from changes in the economy and the family (Christensen 1988a; Huws 1991). As Haraway concludes, "The homework economy as a world capitalist organizational structure is made possible by (not caused by) the new technologies" (1990, 208). Computer work is only one aspect of the range of work that can be, has been, and is being done at home. While telecommunications and information technologies are making it possible for work to be done almost anywhere, how the technologies are used and who has control of the information affect the impact of these technologies on our daily lives.

Another assumption concerns who works at home. The "electronic cottage," as it is portrayed, is primarily an upper-middle-class phenomenon for people with financial resources and flexibility in employment (Toffler 1980). This image does not reveal the diversity of the home-based work population; increasingly, low-paid homeworkers are being hired on a piecework basis in a variety of occupations (Allen and Wolkowitz 1987; Rowbotham 1993).

A third assumption relates to where people work in the home. In contrast to the high-tech electronic cottage with its communications hub at the heart of the home (Johnson 1990), home-based workers have a variety of different environments based on their priorities and resources. A surprising number are mobile workers, often having several work locations other than their home. People are now working almost anywhere: at home, in clients' offices, even in their cars.

A fourth assumption is that home-based work allows more balance between home and work life. For some that might be the case, but for most, work takes precedence over other facets both temporally and spatially. Especially for women, separating home and work life becomes problematic, as they must cope with the responsibilities of child care, household maintenance, and paid work activities (Christensen 1988b; Duxbury, Higgins, and Mills 1992). In dissecting these four assumptions, differences in work status, gender, and economic class have to be analyzed. These variables affect the choices homeworkers make, the type of work they do, the locations they work in, and the amount of control they have over their time and resources.

Home, Work, and Urban Life

When the home becomes a workplace, the boundaries between work/home life, workplace/home, public/private space, and male/female roles become blurred. How these separate spheres and roles are interpreted by people working at home affects the home as a social and physical setting. Since the Industrial Revolution, the home in Western industrial society has been defined in opposition to work as a refuge from the public world (Saegert and Winkel 1980). The home in late-twentieth-century North America provides both security and a focus for the identity of the nuclear family. It functions as a synthesizer of experiences, filtering out the uncertainties of the outside world and temporarily providing an atmosphere of well-being. It has also come to be a symbol of self-identity (Cooper Marcus 1995). Increasingly, however, for the urban household the home is becoming little more than a place to sleep, eat an occasional meal, and store personal possessions, as most waking hours are spent elsewhere. Home-based work is changing that pattern.

Though there have been few comprehensive studies that investigate the impact of telework, futurist writers believe that through homework using information and telecommunications technologies, the home is reemerging as a central unit in society with enhanced economic, educational, and social functions (Toffler 1980; Toffler and Toffler 1995; Naisbet 1982; Naisbitt and Aburdene 1990). These writers contend that powerful forces are converging to promote a massive shift of jobs out of factories and offices, and back to the home. They describe a decentralized economy based on the use of telecommunication technologies to produce and consume goods and services. Furthermore, the electronic cottage will foster entrepreneurship and self-employment, resulting in smaller corporations and new organizational structures. Moreover, because the creation of separate spheres for work and domestic activities after the Industrial Revolution was one of the causes of the breakdown of the extended family, the reversal of this separation is predicted to strengthen the family. It is believed that the electronic cottage will lead to greater flexibility in roles and more communality among family members. However, the futurists' rosy predictions are countered by critics with concerns about telework.

Just as information and telecommunication technologies can allow households greater freedom to select their community resources and interactions, these technologies can be used to achieve greater social control. In his writing on disciplinary power, Foucault described the role of the built environment as an instrument of power relations (Rabinow 1984). Though the modernist concept of transparent, flowing space and the use of large expanses of glass cannot be directly attributed to this control function, the transparency and open plan of office buildings and factories does allow for a high level of surveillance. New information technologies free surveillance from the limitations of direct vision; video cameras, telephones, and computers in the new "smart" buildings allow continuous monitoring and surveillance. Workers telecommuting from home can be subject to this same supervision, effectively having their homes turned into "electronic sweat-shops." Home-based activities that transform the household into a sphere for production and consumption are part of a process of economic and social restructuring that opens up new possibilities for flexibility and de-centralization while reinforcing centralized corporate control of the economy. Electronic homework is perceived as "a new frontier in the scientific management of society ... [in which] household activities of all kinds become subordinated to criteria of technological efficiency and rationalization" (Robins and Hepworth 1988, 159).

Within this context, critics of the optimistic visions are concerned that the increased computerization of the home will have dire social and psychological consequences because the range of services that will be able to be produced and consumed in the home will isolate individuals from interactions in the general society. Individuals in these homes will lead increasingly fragmented and isolated lives, ultimately contributing to the disintegration of community life. The household, they assert, cannot be isolated from the social context in which it is embedded (Castells 1985, 1989; Robins and Hepworth 1988; Swift 1995; Rifkin 1995). These critics predict that people will increasingly interact more on the basis of function and interest than propinquity (Webber 1964). Another common concern in the literature is that we are evolving into a bifurcated society of those who will have access to telecommunications and information technologies and those who, because of lack of education, skills, and resources, will not (Castells 1989).

Feminist researchers see certain advantages to telework for those in the population, such as mothers of young children, the elderly, and the physically challenged, who need or want to stay at home (Christensen 1986). Working at home may be one of the few opportunities these people have to earn an income, and they generally like the concept because they have no alternatives. However, there is a danger that home-based work will be viewed as a substitute for child care services. If more businesses provided child care facilities and if more flexibility in work patterns were offered, homework would perhaps not be such a strong option. There is also concern that for many women the nature of homework is exploitative. The trend toward piecework in the home may cause women's working conditions and pay to deteriorate, forcing them into an unending cycle of work on both household tasks and piecework. Specific issues that have been raised regarding telework include financial exploitation of homeworkers, poor conditions of employment, lack of union representation, restrictive residential zoning, and reluctance by management to relinquish control over employees. There are problems also of spatial constraints and conflicts for people who live in small houses or apartments that are quite unsuitable for homework.

Though pressures in society are making home-based work necessary for segments of the population, some critics are concerned that the psychological problems of working at home have been underestimated (Forester 1988). Possible difficulties with working at home include lack of motivation and discipline, inability to organize work and manage time effectively, and problems in being a self-manager. In addition, many consumers have found that new information-based services (such as teleshopping) are not useful and do not fulfil their psychological needs. Homeworkers may have psychological problems that arise in relationships with their family or spouse, and they may have feelings of loneliness or isolation from colleagues and concern about social status, especially in the neighbourhood.

Organized labour opposes telework, arguing that the same issues that surround garment-making apply to information workers (Huws, Korte, and Robinson 1990). Though part-time work at home may be appropriate for managers and some professional employees, unions claim minimum wage and labour standards cannot be guaranteed for clerical and support workers in the home. Unions fear workers could be exploited in electronic sweatshops where they can be electronically monitored and kept in isolation from their colleagues. In an effort to avoid such exploitation, the AFL-CIO and the Service Employees International Union (SEIU) both passed resolutions for a ban on computer homework in their 1983 conventions (DiMartino and Wirth 1990). However, since that time the unions' position has weakened. In 1989, the Eighteenth Constitutional Convention of the AFL-CIO opted for legislative control rather than legislative ban, urging the establishment of appropriate new regulations on homework to prevent exploitation of workers in these settings (Mahfood 1992). In Great Britain, the white-collar trade union MSF (Manufacturing, Science, Finance) emphasizes that it is not opposed to new forms of working but wants changes to traditional working practices to be implemented by negotiation and agreement rather than management decree (Bibby 1999). Despite these efforts, both the public and private sectors are expanding opportunities for people to work at home and in other non-traditional work environments, such as prisons, with little (or no) regulatory control.

The assertion that telework will strengthen family life appears to be fallacious, because the futurists base their ideas on a simplistic vision of the preindustrial European family and its work patterns. This vision portrays the preindustrial family as an harmonious unit in which work and home life were seamlessly intermingled and communities were largely self-sufficient. Scholars of the new social history (Hareven 1977) have dismissed the traditional view of the family as a passive agent that broke down under the impact of industrialization and urbanization. Current research no longer analyzes the family as an unchanging institution, but acknowledges that families have differed historically, having constantly evolved and undergone changes. Discovery of the family's ability to adapt to change has led to speculation that the family itself may have acted as an agent of change, preparing members for new ways of life. The family has never been an utopian retreat from the world; rather, it has been diverse and flexible, and has varied in accordance with social and economic needs.

The romanticized portrayal of the preindustrial European family has further confused the image of contemporary teleworkers by providing a speculative model of home life. This portrayal lacks an understanding of the work done in preindustrial cottage industries and the consequences of that work for the household. While cottage industries have been identified with artisans (usually men), the actual workers (usually women and children) in these industries were paid on a piecework basis and did "putting out" work on various components of the manufacturing process, such as cloth-making for master clothiers (Gregory 1982). Their supply of work depended on their employer, and all members of the family had to contribute for economic survival. For pieceworkers, the family life course offered little opportunity for choice because the family lacked control over their destiny and their household economy. Artisans, in contrast, were self-employed craftsmen who used their own tools and sold their services directly to consumers or wholesalers (Pred 1981). Artisans and their families had considerable flexibility in their everyday lives. The futurists have assumed that the experience of working at home will be similar for all individuals and families. But gender, the kind of work, and the degree of control a homeworker has over its execution affect the diverse experiences of everyday life.

In addition, the idea of the home as a place of nurture, comfort, and leisure within which work will be seamlessly integrated is contradicted by the daily pressures and living circumstances portrayed in current statistics. An estimated 28 million American women are battered by their husbands or partners each year, making the home a far more dangerous place for them than city streets. Rather than a utopian retreat from the world, the home can be a place of personal stresses and dysfunction.

With the shift of men's place of work from the home to the factory after the Industrial Revolution, the sexes were separated by time, space, and socially enforced role expectations. Home-based paid work is seen as a way of reintegrating sex roles and responsibilities. However, unlike those writers who focus on the impact of homework on the individual family and its dwelling, feminist theoreticians (Hayden 1984; Saegert 1980) question the notion of the separate spheres of home and work, and postulate a new paradigm of the home, the neighbourhood, and the city that supports, rather than restricts, the activities of working women and their families.

The role of the home and workplace for at-home workers is changing. In describing the home of the future, futurists have taken a technologically determinist analysis that obscures the role individuals and society play in the process of change. The use of telecommunications and information technologies to generate and transmit information has the potential to decentralize resources and provide flexibility and control over both work and home life. At the same time, such technologies could individualize homeworkers and isolate them from opportunities in the larger society. Many questions regarding this phenomenon need to be addressed. How do workers adapt to the new form of workplace in the home? How does home-based work affect social relations both within the household and in the larger community? How do the activities of a home change and how is this change accommodated physically when people work at home? Generally, is working at home a viable alternative to other forms of work arrangements?

To begin to understand this trend, Chapter 2 describes a typology of flexible workers, and situates this internationally. This analysis is supported by data derived from three studies of teleworkers and home-based entrepreneurs conducted by the author in San Francisco and Sacramento, California, in 1990, Canada-wide in 1995, and in Vancouver in 2000, as well as other research. The research spans a ten-year period in which significant changes have occurred in organizational structures, technological innovations, and family priorities, impacting the way work is and could be conducted. The studies had significant findings regarding the mobility of this workforce, its distinct divisions according to work status and gender, and the tensions encountered in trying to combine work and domestic activities in the same setting. The following chapters amplify these findings and outline their implications for the social and physical environment of urban North America. Chapter 3 articulates the significant themes from the 1990 study and describes patterns of work and home life and psychological profiles of home-based workers using both qualitative and quantitative data. These patterns illustrate that often existing gender roles are not altered in the home, and the new role of paid worker is difficult to accommodate in it.

Chapter 4 describes the findings of the 1995 Canada-wide survey focusing on teleworkers and home-based entrepreneurs and demonstrates that home-based workers can not be treated as a single group. Work status, gender, and economic class affect experiences. In Chapter 5, a case study is developed to analyze the consequences of the networked economy on a particular locale, Vancouver, British Columbia. Findings reveal a bifurcated workforce made up of highly skilled, highly paid knowledge workers and low-skilled, low-paid pieceworkers made redundant by new technological capabilities.

Three interrelated concepts need more careful examination now that working at home could potentially alter how homes and neighbourhoods function: home, community, and sense of place. One of the recurring tensions in North American society has been between the values of independence and individualism, and the values of community linked to the desire for connection and caring. The home in relation to its community setting has become the physical manifestation of the tensions that are occurring in the larger society. For home-based workers these stresses have become exacerbated. Chapter 6 examines several paradoxes in the relationship between home and work. Instead of promoting a natural wholeness to everyday life in which work and home life become a seamless fabric, the polarities between these two spheres are even more difficult to resolve when working at home.

Before the Industrial Revolution, working at home was the norm. Once the live/work relationship was severed, for most people, the home was redefined as a refuge from public life. Teleworkers must now carry out their activities in a home setting that does not support their new identity, and one that is vastly different from what constituted home for the cottage workers of the Middle Ages. While it may appear that home-based workers should have a stronger attachment to their immediate locale, the neighbourhood, than those who go elsewhere to work, this is not borne out by empirical evidence. As the studies presented in Chapter 7 demonstrate, teleworkers rarely use their neighbourhoods, especially in those neighbourhoods where there are few services and people during the day. Until there is a critical mass of homeworkers, and services supporting them, the neighbourhood will hold few opportunities for social contacts.

Correspondingly, home-based work activity raises questions about the nature of community. Some teleworkers, disconnected from their neighbourhoods and the social networks of office life, rely on computer networks for their social lives. These networks change their perception of their sense of place in the world, opening up opportunities for geographically wide exchanges, yet limiting these exchanges to those that are electronically mediated. As articulated in Chapter 7, information and telecommunications technologies are capable of creating "virtual" workplaces and communities in the privacy of the home. For people working at home using these technologies, the home is being transformed into an "information factory" where work can be created, processed, and disseminated, eliminating many workplace-related functions. Work in these home settings takes precedence over home activities both spatially and temporally.

The home for teleworkers is no longer a place of refuge, since work-related stresses become associated with the home. Moreover, though the home for home-based workers becomes more insular, it is no longer a buffer from the complexity of urban society. Individuals working in these homes may be isolated from the world outside their doors, and from other workers, but are inextricably linked to the global society. The concluding chapter makes a plea that the trend toward the atomization of work, home, and community life be treated as a societal issue, rather than being portrayed solely as a matter of individual choice.

While the themes presented here might seem disparate, on closer examination they reveal a holistic portrait of the trend toward flexible work arrangements, their effect on people's daily lives, and the corresponding impact on urban patterns in homes and communities. The following chapters document the tensions between domestic and work life that are manifested when work is conducted at home: between the desire for flexibility and the tendency for work to become "out of control"; and between freedom and invisibility and isolation. The role of technology in precipitating these tensions becomes significant. The social and spatial relations that emerge describe a dispersal of activities away from traditional nodes and forms, a reconcentration in other nodes in other ways, and a polarization and disparity based on gender, class, and work status.

2
Profiling the Teleworker: Contextualizing Telework

Determining the magnitude of telework is a problematic endeavour. Accurate statistics are hard to arrive at because telework is often included with statistics on other forms of home-based work, and both terms are inconsistently defined. To understand the extent of home-based work, distinctions have to be made between the various categories of work organization. Treating them as one category presumes that all home-based workers have similar priorities and resources. Home-based work can be defined as paid work conducted in the home or from the home, on either a part- or full-time basis, though some workers, such as employed teleworkers, often have another workplace in the organization in which they work. Home-based workers can be self-employed or corporate-employed and they may or may not use telecommunications equipment. Their work includes most types of work found in the general society.

Home-based work encompasses a significant portion of the working population in North America and is a growing trend. Ten years ago in the United States, according to the May 1991 current population survey (CPS), approximately 20 million non-farm employees were engaged in some work at home as part of their primary jobs, representing 18.3 percent of the total who do paid work (Deming 1994). Of those who worked at home, more than 60 percent brought work home from their workplace and were not paid specifically for that work. Of the rest, 5.6 million people were self-employed. Wage and salary workers who were paid for hours worked at home accounted for 1.9 million, or less than 10 percent of the 20 million who did any work at home. Using a broader definition of home-based work, a US Department of Transportation study using data from LINK Resources Corporation, a market research firm specializing in telecommuting, estimated the population of at-home workers at 38 million people or 30 percent of the US labour force (US Department of Transportation 1993). The majority of those were self-employed or brought work home after regular hours. Of the rest, 2 million workers could be typified as telecommuters. Helling (2000), using the

1995 US National Personal Transportation Survey (NPTS), presents more variegated data. Her study differentiates between home workers (the 5.7 percent of total workers who give their primary workplace as "at or out of home"), telecommuters (the 8.8 percent who give a non-home primary workplace and work at home at least eight hours every two weeks), home-based mobile workers (0.1 percent who give their primary workplace as "at home or out of home" but drive more than ten miles on a given day as part of work), and non-home-based mobile workers (2.1 percent who report that they have "no fixed (primary) workplace"). This amounts to 16.7 percent of total workers, or 22,118,602 out of a total US worker population of 131,697,367.

The 1992 national survey conducted for the Canadian Home-Based Business Project Committee defines the estimated upper limits for homework in Canada (Orser and Foster 1992). Including all types of home-based workers from the full-time home-based business operator to the office worker bringing work home, the study estimated that up to 23 percent of working Canadians, or 2.17 million of the workforce, spent at least some of their working time at home. Of those, 38 percent were supplementers (i.e., an employee who brings work home), 25 percent were self-employed full time, 23 percent were self-employed part time, and 14 percent were substituters (i.e., an employee who spends the workday at home). Using a more restrictive definition, the 1991 census estimated that 1.1 million Canadians used their home as their usual place of work, of whom a quarter were farmers (Statistics Canada 1991a). The 1991 Survey of Work Arrangements estimated that approximately 600,000, or 6 percent of employed paid Canadians, worked all or some of their regularly scheduled hours at home (Statistics Canada 1991b). The 1996 census revealed that the percentage (6.1 percent) of the home-based work population had not grown in proportion to the whole workforce but had increased in numbers to 1 million (Statistics Canada 1997).

While these varying statistics reveal the trend they also amplify the difficulty of arriving at an accurate estimate of the total number of home-based workers, because definitions vary of what constitutes a home-based worker. The lack of accepted definitions for people who do paid work at home reflects the invisibility and occupational variability of homeworkers. Many home-based workers are part of the informal or underground economy. They are reluctant to reveal that they work at home because they do not want to report their earnings or be found in violation of local zoning ordinances. Often statistics group all home-based workers without differentiating between full- and part-time workers, nor the corporate-employed and self-employed. A closer investigation of the above estimates divulges that a significant portion of home-based workers are in professions that have long required some work to be conducted from home: sales, insurance, real estate, or teaching.

Most surveys on home-based work are designed to cast a wide net. Job-related, income-producing work at home can range from a salaried employee working out of a briefcase at the kitchen table at night to a full-fledged home-based business. There are also problems in trying to formulate "employed" and "self-employed" categories for this non-traditional workforce; the same person who spends time working at home for an employer may also be moonlighting in his or her own business and working on a contract for another company. The large number of estimated at-home workers in some reports may also be based on questionable methodology that poorly defines and interprets phrases like "income-producing" and "job-related." To further complicate the statistical analysis, home-based workers are often grouped with the self-employed, who may or may not be homeworkers. Though the rate of self-employment has steadily declined since the nineteenth century, this trend has been reversing in the past twenty years. Statistics on teleworkers are equally difficult to arrive at as there are few formal telework programs compared to the number of informal arrangements.

Nevertheless, both the self-employed home-based entrepreneur and the corporate-employed teleworker are part of a growing phenomenon (Pratt 1993). An analysis of employment patterns reveals that home-based work has grown over the last two decades but has levelled off in the last few years. In Canada, the 1996 census found that just over 6.1 percent of the employed labour force in census metropolitan areas worked at home (Statistics Canada 1997), compared to 3 percent in the 1981 census (Nawodny 1996). In the Greater Vancouver Regional District homeworkers increased from 3.9 percent of the employed labour force in 1971 to 7.2 percent in 1991 (Baxter 1994) and 8.2 percent in 1996.

The 1996 Canadian census also revealed distinct patterns of those working at home in terms of gender, age, work status, and socioeconomic level (Statistics Canada 1997). Of the 818,625 people working at home in non-farming occupations, 54 percent were female, and of those women, 55 percent were between the ages of thirty-five and fifty-five, with 22 percent between twenty-five and thirty-four years old and 12 percent between fifty-five and sixty-four. Just over half (52 percent) of the women were self-employed, while 44 percent were paid workers and 4 percent were unpaid family workers. As well, just over half (51 percent) worked full time, while the rest worked part time. The predominant occupational categories for women were sales and service occupations (36 percent) and business, finance, and administrative occupations (33 percent). Of the women working in sales and service occupations, almost half worked as babysitters or nannies, or in early childhood education, and one-fifth worked in sales positions. Of those women working in business, finance, or administration, almost two-thirds were clerks, bookkeepers, and secretaries. Other occupational categories included art, culture, recreation and sport (9 percent), management (7 percent), social science, education, government service and religion (5 percent), and all other occupations (10 percent). The average annual income of female at-home workers was $19,208.

The demographic profile for men working at home in non-farming occupations reveals a disparity between the sexes. Men's average annual income was $31,117, 62 percent greater than women's. Two-thirds of the men were self-employed, 11 percent more than women, while one-third were paid workers and 1 percent unpaid family workers. While only half of the women worked full time, almost three-quarters (73 percent) of the men worked full time. Just over half (52 percent, compared to 55 percent of women) were between the ages of thirty-five and fifty-five, with 16 percent between twenty-five and thirty-four years old and 17 percent between fifty-five and sixty-four. The predominant occupational categories for men were also different. While sales and service was also the largest occupational category, it was a smaller percentage (21 percent) among men. The other categories included trades, transport, and equipment operators (17 percent), management (17 percent), business, finance, and administrative (14 percent), natural and applied sciences (9 percent), and all other occupations (23 percent). In short, men working at home generally have a larger income, mainly are self-employed, work full time rather than part time, and are employed in more management-type occupations. In contrast, women are more likely than men to be paid workers, work part time, and be involved in service occupations.

In the United States and Canada, there is a continual reported rise in the number of telecommuters and remote workers (i.e., those working away from their corporate headquarters in other offices), though varied definitions of "home-based work" and "telework" make it hard to get accurate data. A January 1999 survey documented that there were 11 million personal computers supporting telecommuters in the United States, a significant increase from a 1997 study that documented under 8 million telecommuters (Chun 1999). In Canada, Statistics Canada does not report data on telework separate from home-based work but the 1996 Census did reveal that all forms of home-based work had increased 40 percent from 600,000 workers in 1991 to one million in 1997 (Statistics Canada 1997).

The growth in the number of teleworkers is attributed by many to the high concentration of home computer ownership in Canada and to one of the cheapest Internet access rates in the world (Canadian Telework Association 1999). Statistics Canada found that 4.2 million (or 36 percent) of Canadians owned a home computer in 1997; it estimates that by the year 2000 almost half of Canadian homes will have a home computer. Concurrent with this is Internet use. Results of Statistics Canada's latest survey (1999a) show 4.3 million households (more than one-third of all Canadian households) have someone using the Internet from their home, work, or another location, an increase of 25 percent from 1997. A November 1998 CBC-commissioned study more optimistically showed that Internet-connected homes jumped from 13 percent in 1997 to 23 percent in 1998 (Nielsen Media Research and Canadian Broadcasting Corporation 1998). Canadian Internet use is, however, a fraction of the estimated 163 million users worldwide in 1999.

ICT in an International Context

While the extensive use of information and communication technologies (ICTs) for work, communication, commerce, and even socializing is a significant trend at the beginning of the twenty-first century, it is important to put this phenomenon in the larger international context to understand its significance. The assumption that the emergent information society neatly translates into the knowledge society is spurious, given the fraction of the world's population who have access to computers and associated technologies. Out of the 5 billion people in the world, 4.9 billion are excluded from the so-called "wired" or "network" society because of lack of resources and inadequate infrastructure and expertise (Mitter 1998). These excluded people, rather than relying on ICT to help them process information and make decisions, have developed, through trial and error, locally based science and technologies.

Of those who are "wired," the United States and Canada dominate the use of personal computers, with approximately forty PCs per hundred inhabitants. Australia and New Zealand have around thirty PCs per hundred people, and the European Union around twenty. However, in Latin America and Eastern Europe, there are no more than three PCs per hundred inhabitants, in developing Asia fewer than two, and fewer than one in Africa. Out of the current (September 1999) on-line population (i.e., those connected to the Internet) 57 percent are from North America, 21 percent from Europe, and 17 percent from Asia, while 3 percent are from South America, only 0.75 percent from Africa, and 0.5 percent from the Middle East (NUA Internet Surveys 1999). It is also the United States, Canada, and Europe that account for over 90 percent of Internet hosts. Consequently, the majority (63 percent) converse in English, while the next largest group of users (14 percent) converses in Japanese, followed by 7 percent German, 6 percent Spanish, 4 percent French, 2 percent Korean, 1.5 percent Italian, 0.9 percent Portuguese and Mandarin, 0.4 percent Cantonese, and 0.3 percent Hebrew (Mitter 1998). There are also distinct differences between the sexes in Internet use, which are often predicated on cultural factors that inhibit women's access. In the United States it is estimated that 64 percent of the users are men; in Europe that number rises to 82 percent; and in the rest of the world the estimated number of male users is 76 percent of the total (NUA Internet Surveys 1999).

These statistics reveal a distinct disparity between rich and poor nations in access to telecommunications and information technologies. There is also unequal access within countries to these resources, as the statistics on use by gender reveal. Developing countries often have prohibitive telephone costs that exclude the majority of the population from connectivity. Furthermore, telephone access is mainly limited to urban areas. For example, in Vietnam almost all of the telephone lines are in the five major cities. This further exacerbates the disparity both between rich and poor and between the sexes, as often in rural areas poor women form the majority, having less opportunity to migrate to urban areas.

Political considerations often inhibit the use of new technologies. In China and Vietnam, for example, there are restrictions on who can use the Internet as the political leaders in these countries are concerned about its use to foment dissent. Individuals and groups can make powerful use of telecommunications and information technologies, such as the Web site Mexican revolutionaries in Chiapas used to convey information to the outside world in the 1990s and the fax machines used to send information to media during the 1989 Tiananmen Square demonstrations in China.

Cultural influences also affect the use of ICT. In Japan, a highly "wired" country, Internet access is hampered not just by high access fees but by mid-level corporate management who don't understand the technology and are reluctant to use it. Japanese cultural biases emphasize consensus and affirmation of that consensus through paper contracts. Faxes are still the most common form of communication in the Japanese corporate world (Lazarus 1999). By contrast, in India the prohibitive costs of telephone access and the long tradition of microentrepreneurship have resulted in a proliferation of roadside telephone, fax, and Internet kiosks. Their success is attributed to cost-effective connectivity and to the communal tradition still prevalent in Indian society (Mitter 1998).

Community-based neighbourhood work centres focused on ICT work are being organized. The most developed model, known as the telecottage, is found in remote areas of the United Kingdom. Telecottages offer telework services to local or regional organizations and provide training to customers (Korte and Wynne 1996). Business communication centres, often operated by women, are now found in most countries of the world. For example, women in Accra, Ghana, have set up such services to provide fax, telephone, copy machine, and computer access.

Neighbourhood work centres and satellite offices are also of increasing interest as a substitute for working at home for teleworkers and entrepreneurs (Mokhtarian 1991b). These centres are located within a convenient commuting distance of the majority of employees utilizing the site. Satellite offices of a firm are relatively self-contained divisions physically separated from the parent firm. Neighbourhood work centres are offices equipped and financially supported by different companies or organizations. Self-employed entrepreneurs and employees of different organizations or different divisions in one organization share office space and equipment in a location close to their homes. Because a variety of businesses, including one-person firms, share services, start-up costs for small businesses are reduced. The entrepreneurs and employees in these centres have opportunities for social interaction, hierarchical structures are generally lacking, and supervision of work is carried out remotely. While satellite offices could alleviate stress due to commuting, the possibility that employees at these work sites will become second-class corporate citizens, losing the benefits and opportunities of their parent company, should be recognized. In addition, satellite offices create the potential for the development of the "company town syndrome," if only one corporation's offices are located in a residential setting.

While community-based services have been highly successful in many parts of the world, a pilot project from 1991 to 1997 to develop telecommuting centres in California proved much less effective (Mokhtarian et al. 1997). Supported by CalTrans, the California Department of Transportation, and the Federal Highway Administration, the project evaluated the effectiveness of these centres as an institutional work arrangement and as a strategy to reduce traffic. There was a high attrition rate among users and few of the telecommuting centres are still operating. Some of this lack of interest can be attributed to socioeconomic and cultural factors. The highly mobile American workforce, often lacking community ties, is more likely to want alternative workspaces to be located in the home than in the community.

Nevertheless, distant work provides potentially significant opportunities. The demand for increased connectivity through ICT is worldwide. Pockets of innovation could be developed anywhere with sufficient bandwidth. The market for ICT services will require customizing for particular locales and cultures, therefore requiring local expertise. However, the regional disparities in cost and quality of infrastructure, language, regulatory climate, organizational cultures, and basic skills of workers and managers may make the development of an ICT sector in a particular country difficult.

Of critical importance is the issue of training so that workers internationally (especially women) will not be relegated to low-skilled and low-paying ICT work such as data entry. Transnational corporations have tended to offload this work to countries with low wages and a pliant labour pool. The movement of large corporations to outsource data entry and other routinized functions to developing countries has to be viewed with caution. While the imperative for this move is to keep labour costs down, the advancement of technologies to allow image processing and voice recognition, among other changes, may make these skills redundant. The ICT employment sector is following a pattern found in other sectors: research and development is located in developed countries, and outsourced work is either low-skilled or highly specialized, thereby impeding the development of a comprehensive ICT sector in a developing country.

If local concerns can be combined with ICT, a myriad of different relationships to the technologies and their uses are possible globally. Because of the constraints of socioeconomic and cultural realities, however, the tendencies now are toward disparities between countries and within countries based on sex and economic class. In this "network" society, those who have control of ICT are the "haves" and those who don't or who rely on others for access and work opportunities are the "have nots." This is now being termed the "digital divide" (Chu 2000; Waddell 1999).

The Evolution of Telework

Several dominant themes emerge from the trends of telework and home-based employment. One of the critical issues concerns the cause of telework. While the computer is portrayed in the popular press as driving this trend, this analysis is a simplistic and technologically deterministic approach that ignores the complex relationship between technological development and social and spatial processes. The diversity of the teleworker population has not been incorporated into this analysis, which assumes that all teleworkers' experiences are the same.

Huws (1991) contends that those who support telework tend to discuss the advantages of self-employment and homework from personal experiences, without empirical evidence for their conclusions. These descriptions of working at home are largely drawn from middle-class experience. Most of those writing about home-based work are technical/professional and most computer owners are middle- to upper-income. Because of this myopic vision, financially comfortable baby boomers are targeted as the group most likely to be working at home (Analysis 1988). This group is concerned about such quality-of-life issues as priority changes to allow family life to predominate over work life and the creation of additional possibilities for women to work after having a family. The increase in the number of homeworkers is attributed to dual-earner, highly educated professionals seeking to avoid the toll that their stressful work life takes on them and their families (O'Hara 1994).

There has long been a tradition of home-based professionals in North America and elsewhere. Professors, scholars, writers, craftspeople, and artists have used the solitude of the home to do most of their intellectual and creative work. Some lawyers, doctors, veterinarians, architects, and accountants have opted to maintain their offices at home. After the Industrial Revolution, however, the need for supervision, communication, and the cooperative use of resources and equipment generally led to the centralization of the workplace in factories and offices. Smaller, cheaper computing and telecommunications technologies have now made more kinds of work portable, creating the possibility for a variety of flexible work arrangements.

This perspective, however, is one of only many when analyzing the trend of home-based work, and especially telework. Huws emphasizes her contention "that the 'electronic homeworker' has become a highly charged symbol, embodying for many their hopes and fears about the future of work. However, the meanings it carries are not constant. Not only have they changed over time; they also vary according to how their holders are placed in relation to the technology, to their work and to their homes – whether, for instance they are men or women, employees or employers, living alone or caring for others, well or poorly housed, young or old, attracted to information technology (IT) or repelled by it" (Huws 1991, 20).

First seen as an energy saviour in the 1970s (Nilles et al. 1976; Harkness 1977), telework and home-based employment were also linked with the recognized need to humanize corporations that was the product of the 1960s youth rebellion (Bell 1973; Schumacher 1973). These portrayals of decentralized work options were implicitly male, middle-class, and highly individualistic, with the freedom to choose work style for greater personal fulfilment. Toffler (1980) encapsulated this perspective in the "electronic cottage." Predictions were made about the "home of the future," which would utilize the latest information technologies to allow people to conduct all aspects of their lives in the sanctity of the home, surrounded by a warm and loving family (Mason et al. 1984).

With the widespread introduction of computers into corporations in the 1980s, their use spread from the sole domain of managers and professionals to low-paid secretarial and clerical staff, who were predominantly women. Concurrent with this development was the perception of the computer as an instrument of control that would increase accuracy and productivity, and allow monitoring. In this context, telework brings not new freedom, but isolation, atomization, and exploitation (Huws, Korte, and Robinson 1990). Concurring with feminist writers, this perspective sees the home not as a haven but as a site of oppression (Oakley 1974).

Countering these arguments were undocumented assertions that homework allowed for greater integration of home and work life, allowing workers (read "women") to both care for their families and still be productive members of the paid labour force (Aldrich 1982). Studies on the experiences of homeworkers have not borne out these assertions (Olson 1983; Kawakami 1983). While some have successfully manoeuvred between work and domestic activities, many dislike the isolation, insecurity, and constant demands that working at home entails.

In the late 1980s the arguments supporting telework and home-based employment became more utilitarian. Corporations, recognizing that to survive they had to be leaner, were looking for a way to reduce overhead costs and increase organizational adaptability (Atkinson 1984). A flexible workforce, based not in a corporate headquarters, but at home or in some other alternative work site, was the answer. Coupled with this development was governmental recognition of entrepreneurship and self-employment as a viable strategy in a stagnant economy. In addition, the young people entering the workforce in the 1980s had been raised in a society with shifting economic and social values and for many, secure employment was an unattainable dream. Flexible employment, such as contract work, is still what most get initially. Concurrent with the economic reality of limited expectations are shifting social expectations among some of the population, recognizing sustainability as an important goal. Telework, by allowing the efficient use of urban space (e.g., live/work) and reducing the consumption of material and energy resources (e.g., less travel for work), fits into the framework for urban sustainability.

The 1990s and now the twenty-first century have seen a significant increase in work that is conducted entirely on-line, from Web site designers, computer graphic artists, systems analysts, and programmers to on-line stock traders. These teleworkers, while well paid when they are working, often need to get and maintain contracts to keep their incomes stable. Fluctuations in the economy and corporate priorities directly impact their work prospects. Telework has been variously interpreted in the past. Currently, the impetus for flexible employment, driven by economic imperatives, that makes telework attractive as a survival mechanism has also created a far less secure workforce.

Profiles

Telework, defined as work-related substitution of telecommunications and related information technologies for travel, encompasses a broad spectrum of work arrangements from full-time employment to contract work and self-employed entrepreneurship. While employed teleworkers would fit into the category of those with secure employment, independent contractors and call centre workers are part of a vulnerable labour force with few benefits and little long-term security. In general, teleworkers cannot be treated as a single group. Work status, gender, and economic class, as well as housing, neighbourhood, and regional differences, affect the choices they make and the resources they have available. Programs and policies addressing home-based work should recognize these differences and incorporate solutions that are specific to particular segments of the homeworker population. In particular, teleworkers, independent contractors, and home-based entrepreneurs should be treated as separate groups. Table 2.1 outlines the types of home-based workers, profiles of which follow.

Table 2.1

Typology of home-based workers

Work status	%	Definition
Employed teleworker/ homeworker/ telecommuter	3	Someone who works away from an employer's office or production facility, often at home, part or full time, communicating via telecommunications technologies as an employee for a public institution or private corporation
Independent contractor	11	Someone who works from home, part or full time, as a contract employee or piece worker on contract to one company
Self-employed consultant and home-based entrepreneur/ business operator	48	Someone who works from home, part or full time, doing consulting work for more than one company or individual; someone who works from home, part or full time, providing a service or product to a variety of clients or customers
Moonlighter	Included in above three categories	Someone who works from home part time as a supplement to a primary job
Occasional homeworker	38	Someone who brings work home after work hours from a workplace on a frequent to occasional basis

Note: Data on the breakdown of home-based workers were derived from Orser and Foster (1992) and Statistics Canada (1991b).
Source: Reformatted from Gurstein (1995).

Employed Teleworkers

Teleworkers, who are often highly skilled and employed, rather than on contract, are primarily upper-middle-class professionals with financial resources and flexibility in employment. These teleworkers work away from the employer's office or production facility, often at home, using telecommunications and information technologies to communicate with their offices. Their main reason for choosing such a work situation (unless they have been pushed by their employer) is work flexibility, and they appreciate the control over their time that working at home affords them. Most of these employed teleworkers spend part of the week at home and part in the office, thereby avoiding such problems as isolation, loneliness, and invisibility. Working in both locations also helps lessen the concerns of managers and office-bound colleagues. It is anticipated that the telework population will grow to include non-managerial and professional employees, including support staff. This growth, however, will require organizational change in corporations regarding performance evaluation and managerial control (Sundstrom 1986).

Teleworker employees, while garnering the most media attention, are still a very small segment of the total home-based work population, though exact numbers are hard to determine. A recent report concluded that telecommuting in the United States was practiced by approximately 2 million workers in 1992 (only 1.6 percent of the total US labour force) but it could reach 7.5 to 15 million within a decade (US Department of Transportation 1993). A 1997 study from Telecommute America, an organization formed to promote telecommuting, also found that 62 percent of 500 companies surveyed have more employees working at home than they did two years ago. The survey defined a telecommuter as an employee who works at least one day a week at home, a satellite office, or while on the road, keeping in touch via computer (Telecommuters 1997). In Canada in 1992, approximately 300,000 workers (3.2 percent of the total labour force) were classified as employees who spend the workday at home (Orser and Foster 1992). This figure may not have increased much since, given that many of the telework pilot programs instituted in the early 1990s have ended due to changing economic conditions and downsizing. In the United Kingdom, a study suggests that 6 percent of organizations are involved in teleworking but under 0.5 percent of the labour force can be classed as electronic homeworkers (Huws 1993).

The gendered nature of telework has not been fully articulated but the 1995 Canadian survey on telework and home-based employment incorporated into this book did find that 61 percent of the public sector teleworkers (i.e., those who work for a public institution or Crown corporation) and 47 percent of the private sector teleworkers (i.e., those who work for a private corporation) were female. In contrast, 81 percent of the independent contractors and 58 percent of home-based business operators were female, while only 38 percent of the self-employed consultants were female. Women and men seem to participate in telework programs at comparable rates. The presence of more women in public sector telework programs is primarily due to their strong presence in public sector employment.

Independent Contractors or Self-Employed Subcontractors

As described in the introductory chapter, low-paid homeworkers are being hired as pieceworkers in a variety of occupations from data processing to garment making. These workers are predominantly women, and typically hired on a part-time or temporary basis. They are called independent contractors because they are not on a regular payroll and work on a contract or piece-rate basis. However, they are often treated as employees since they work for only one company, work on materials provided by the company, and are often directly supervised by the company. Nevertheless, when they work at home they have no guarantee of regular hours, no employee benefits, and few opportunities for advancement within the company.

Industrial homeworkers who assemble clothing, textiles, electronics, toys, and other products that would otherwise be manufactured in a factory are a subset of independent contractors who often are part of the underground economy and have limited resources and skills. It is impossible to determine the extent of industrial homework, much of which is invisible, but industrial homework is reemerging on a significant scale in the cities of the industrialized and developing world (ILO 1995). There are close to 1 million industrial homeworkers in the United Kingdom (Rowbotham 1993) and as many as 100,000 in Canada (Ontario District Council 1993). Mitter (1986) has described the growth of industrial homework as the "creation of the Third World within the First World." Studies done throughout the world (Rowbotham and Mitter 1994; Mitter 1992) have arrived at similar conclusions: industrial homeworkers, most of whom are low-income women, are especially vulnerable to inadequate working conditions and substandard working environments.

Skilled professionals who have lost their jobs as a result of corporate restructuring are now being found in similar contractual relationships. As organizations strive to become more flexible to meet changing market conditions, there is an emphasis on downsizing (now called "rightsizing") and reorganization of the corporate structure or "reengineering" (Drucker 1988; Hammer 1990). Part of what is being called the "flattening out" of an organization or "dis-intermediation" is the loss of middle managers and support staff (Tapscott and Caston 1993). While many of these jobs are permanently lost, some former employees are being hired back as independent contractors based in their homes or at mobile workstations such as in client offices (ILO 1990).

These "flexiplaces" or "nomadic workstations" are seen by many large organizations as a way to create a more responsive, productive workforce closer to clients and to reduce real estate expenses for corporate headquarters (Joice 1993). Rather than large centralized corporate offices with individual office spaces and cubicles, the new direction in office planning is toward "hotelling," where a floor of workstations, with one or two support staff to maintain the corporate identity, is shared by those who happen to be in the office that day. Both employees and independent contractors are encouraged to be "footloose," independent of corporate headquarters, and self-reliant in terms of work equipment. When independent contractors establish home offices with minimal contact with superiors and coworkers, the responsibility for buying and maintaining work-related furnishings and equipment is borne by them, not their organizations.

Home-Based Entrepreneurs

While corporate telework programs are garnering a lot of media attention, the largest percentage of home-based workers are entrepreneurs who are either self-employed consultants or home-based business operators. Self-employment is the fastest growing segment of the labour force (Orser and Foster 1992). In the United States, 5.6 million, or 28 percent of the total labour force, and in Canada 1.09 million (including farmers), or 8 percent of the total labour force, are self-employed workers based at home (Deming 1994; Statistics Canada 1997). Of Canadians, 52 percent are self-employed full time and 48 percent part time.

Home-based entrepreneurs often provide the same types of services as independent contractors, but they have multiple clients and contracting arrangements. They have the autonomy to set their own rates according to what the market will bear, and they solicit their work and monitor their own progress. They find the home to be a desirable base for initiating such a venture because the home office can be written off taxes and monthly business expenses can be minimized. Fifty percent of those currently involved in home-based businesses are in a service industry, and studies have demonstrated that working full time they receive about 68 percent of the income of corporate-employed workers (Orser and Foster 1992). The Orser and Foster study also documented that women entrepreneurs are the fastest-growing segment of new small businesses, many of which are started in the home.

The significant percentage of women entrepreneurs worldwide has been documented in numerous studies and has become a rationale for the introduction of microlending practices to assist their businesses (Servon 1995). In many countries women (especially low-income women) participate in the informal economy in proportionately higher numbers than in the formal economy (Berger 1989). Due to domestic responsibilities and sexual stereotyping, which create barriers to formal employment, many women work part time or in the home. Microenterprise, particularly petty commerce, is a common choice. However, since much of women's paid and unpaid work is not counted by government statistics, it is hard to determine the exact level of women's economic activity.

Moonlighters

Another numerically significant category of home-based worker is "moonlighters." Like home-based entrepreneurs, moonlighters provide services or products to a variety of clients or customers, but they are doing this work for supplemental income in addition to their primary employment. While it is difficult to arrive at accurate statistics on the number of moonlighters, as many do not report incomes for this type of economic activity, the CPS (current population survey) in the United States estimated that up to 7.1

million people worked at a second job in May 1991, with one-third of those doing at least some paid work at home (Deming 1994). Of those, it is estimated that a disproportionate percentage are women.

Occasional Homeworkers

It is estimated that 12.8 million Americans – 60 percent of at-home workers according to Deming (1994) or 30 percent according to US Department of Transportation (1993) – bring work home from their workplaces after regular work hours. As well, close to 825,000 Canadians, 38 percent of those who work at home, are occasional homeworkers (Orser and Foster 1992). These workers are not paid for work they do at home. While academics, for example, have traditionally brought work home, this phenomenon is increasing in other sectors as much work now focuses on the creation, distribution, or use of information.

Back-Office or Call Centre Workers

In addition to home-based or flexible workers, back offices have been created in a variety of North American locales away from the central corporate headquarters. Many such offices have been developed in suburban communities to tap the large number of educated female suburban homemakers who want to return to work (Baran 1985). These back offices are primarily data processing centres for such data as insurance claims. The kind of work that is located in these centres rarely offers opportunities to rise up the corporate ladder, the supervisory structure of the corporation is not altered, and the workers are often independent contractors.

Call centres have been developed in a variety of locales for market research and the provision of such toll-free telephone services as direct-order buying, taxi and food ordering, and credit card services. These services can be provided from a home with special telecommunications installed or, more often, from a small office that can be used by a number of people. Call centres servicing North American customers are located in rural areas, in prisons and convents, as well as offshore in the Caribbean. Like back offices, these centres are predominantly staffed by women who work on contract on an hourly or piece-rate basis. A study of call centre workers across Canada found that an average of 70 to 72 percent were women and that a high proportion of call centre employees are youth (under twenty-nine years old) (Buchanan and Koch-Schulte 2000).

The Gendered Nature of At-Home Work

When analyzing the changing perceptions and attitudes toward telework and home-based employment, existing research demonstrates patterns in home-based work activities based on differing experiences between the sexes, economic classes, and work status (i.e., teleworker, independent contractor, home-based entrepreneur). The experiences of home-based work for independent contractors paid on a piecework or hourly basis appear to be distinctly different from those of the self-employed. For example, data processors, most of whom are women, are typically hired as independent contractors on a part-time or temporary basis. Since they are paid on a piecework basis and must fill quotas, their long work hours often interfere with their family responsibilities. Usually, because of spatial constraints, they work in spaces intended for other uses. When they work at home they operate entirely out of the mainstream of the company, often in isolation. In contrast, home-based entrepreneurs, because they are more autonomous, find home an advantageous base from which to expand. Though many women are now starting their own businesses at home, few men are opting to work as independent contractors unless corporate downsizing necessitates it. However, even in the same work, male and female experiences of working at home appear to be distinctly different. Female homeworkers have the dual responsibilities of paid work and family, while male homeworkers primarily view themselves as engaged only in paid work. These variables, as well as other factors such as regional differences, influence the choices made by homeworkers and the resources available to them. In turn, these choices and resources affect decisions about the type of work done, and how and where it is conducted.

The few large-scale studies on home-based work have revealed the fallacy of contemporary discussions of homework that focus exclusively on notions of the "electronic cottage" and "telecommuting." The implicit assumption made by advocates of telecommuting is that the computer determines the decision to work at home and that at-home work cannot proceed without it. This has not been substantiated by research. Over a decade ago, a survey of over 7,000 female homeworkers and in-depth interviews with seventy-five home-based clerical workers found that the reasons to work at home were not technological but related to family responsibilities and the need for income and achievement outside of the homemaker role (Christensen 1986). These findings, however, would probably be different today, especially in the reported use of computers, because much has changed in terms of technological capabilities and cost in the last decade. In contrast to the experiences of low-paid pieceworkers in a survey of 104 professional homeworkers in a variety of occupations from five different cities, almost all were found to have the use of a computer at home (Ahrentzen 1987).

The vast majority of these clerical workers were married women who were hired by companies as independent contractors, not salaried staff, and paid by piece or hourly rates. Christensen's survey found that women laud paid work at home because they can control when and how they work, but express resentment at having their schedules controlled by their families and employers. Major disadvantages were seen to be isolation and the lack of credibility regarding their working status. Working at home requires careful time scheduling, as it has been found that homeworkers can not work and care for children at the same time. Home-based work also requires "a place of one's own" that offers privacy but is not so separate that other activities cannot be supervised. Family support is needed, especially from the partner, to help with child care and household tasks. The study found that employment status is a critical variable in understanding the realities of homework. Current corporate practices regarding independent contracting run the risk of creating a vulnerable second-class corporate citizenry that will be exploited. In contrast, efforts to promote self-employment and entrepreneurship can enhance independence and flexibility for the worker.

Subsequent studies have corroborated many of Christensen's findings. A case study of twenty-six clerical homeworkers at an insurance company found that the women initially welcomed the opportunity to combine wage-earning with household and child care responsibilities, but many soon found that the day-to-day requirements of homework undermined their ability to meet their family responsibilities, creating conflict and stress (Costello 1988). The women felt isolated, trapped, overworked, and underpaid, and they resented that their time was defined in response to the demands of their family or employer. A critical issue was their work's "invisibility," as outsiders presumed that they weren't really working since they were always at home. The women also found that adapting their home to the requirements of homework presented problems related to the lack of appropriate space and intrusion of the workspace into the space of family life. A study that compared 106 home-based and 260 office-based clerical workers presented similar findings (Gerson and Kraut 1988). The researchers found that the work of homeworkers is spread out over more of the day than that of office workers, and that homeworkers are responsible for a higher proportion of household and child care duties than their office counterparts.

Research on other homework occupations further corroborates the findings on clerical homework. Gringeri's study (1996) of seventy-five at-home female assembly workers found that the most compelling reason to be at home was family responsibilities. As pieceworkers assembling automobile parts, the women generally exceeded in-shop production, were highly motivated, worked well without supervision, and frequently were assisted by other family members. Research on "cottage assemblers" in the San Francisco Bay area found that electronics manufacturing in the home changed the nature of relationships there (Lozano 1989). The household, its members and resources were placed in the service of corporate production. In addition to the person who had been contracted by the firm, family and friends provided labour in the manufacturing of electronic components. A study of thirty industrial homeworkers in Toronto, Canada, reported wage and labour exploitation (Dagg and Fudge 1992). The majority of those interviewed were Chinese-speaking, most were paid less than the minimum wage, half reported difficulties in being remunerated for work completed, and many were being helped in their work by their children.

Studies that focused on how the worlds of paid work, unpaid work, and family intersect in the home context revealed the gendered experiences of homeworkers (Olson 1983; Gerson and Kraut 1988). The studies found that the reasons people chose to work at home influenced their experience of homework. Women worked at home primarily to be with their children, while men started working at home for economic and practical reasons to do with their businesses. Neither sex preferred to intersperse work and family activities. Although women had to because of their dual responsibilities, men tended to have the social and environmental support to separate work and family activities. It was found that women view telework as a compromise because of their need for flexible scheduling, while men see it as providing autonomy and as a bonus that allows more interaction with their families.

Further studies reveal that working at home does not change gender roles and that the home remains a sex-segregated environment where women's and men's workspaces reflect their different roles (Wikström, Lindén, and Michelson 1998; Miraftab 1996). While men tended to have defined workspaces that were separated from the rest of the home by a door, women tended to do most of their work in central areas of the home, primarily the kitchen. In addition, working at home does not change the division of labour within the home. There is no fundamental shift in attitudes or practices, although men are doing slightly more housework and child care when they are at home. Nevertheless, men see themselves as primarily working at home; women are torn between their work and family responsibilities. Further research, however, is needed to determine significant trends regarding the involvement of male homeworkers with family activities. Some male homeworkers, however, are either choosing to work at home because they want to be more involved in family life or finding that they are more involved once they start working at home.

Contrary to the idealized vision of homework allowing for a greater balance between work and family life, many of those who work at home work long hours with little time to devote to their families and housework. Besides working during the day, a considerable number of homeworkers also work at night and in the early morning hours (Ahrentzen 1987). This is corroborated by a survey of 10,000 readers of two computer magazines. Of those who did some work from home, 58 percent said their home-working hours were additional to normal working hours (Telecommuting 1987). This trend of "overwork" creates situations in which people work all the time to the detriment of other aspects of their lives. In their investigation of how professionals negotiate the heavy workload required of them, Seron and

Ferris (1995) conclude that it is untenable without a partner at home to manage domestic responsibilities.

Rather than lessening work-related stress, homework actually increases stress for many people, especially women. Homeworkers find they have to work around the time schedule of family members, and because of the lack of spatial separation they are constantly accessible. Another identified problem is isolation. Working alone at home can cause "cabin fever" and low self-esteem. This isolation is similar to the experience of those who are primarily involved in child care and home care. Homeworkers perceive the issue of control in both advantageous and disadvantageous terms. Though working at home may allow some people to control when, how, and where they work, the mingling of home and work life also creates its own space and time conflicts.

Independent contractors, and especially industrial homeworkers, are not part of the romanticized media portrayal of the homeworker as a middle-class teleworker or consultant working out of a "home office" or an artisan making crafts for discretionary income. Instead, industrial homeworkers represent a labour market characterized by a feminization of the workforce, extremely low wages, irregular work loads, inadequate and often dangerous work conditions, virtually no protection of rights or access to social benefits, and the double burden of paid work and household responsibilities including child care (Rowbotham and Mitter 1994; ILO 1990; Christensen 1988a; Gannagé 1986). These conditions are largely attributable to the weakness of the homeworker's bargaining power due to low socioeconomic status linked to class, gender, and ethnicity, combined with the growing "casualization" of work (associated with the "floating" or "flexible" labour force) within global economic restructuring (Phizacklea and Wolkowitz 1995; Rowbotham and Mitter 1994; Leach 1993). For this population of at-home workers, ethnicity can compound their vulnerability due to discrimination from the dominant culture, language barriers, and patriarchal practices within their own cultures.

While clearly teleworkers have significant advantages over industrial homeworkers in terms of flexibility and control over their time and resources, there is a very real danger that this group could encounter conditions similar to those of industrial homeworkers, especially as they become vulnerable to a change in status to independent contractors when they work at home. However, the evidence suggests that highly skilled information workers are at an advantage in the flexible labour market, as they are portable and can easily move employment (Carnoy, Castells, and Benner 1997) while the (predominantly female) low-skilled information workers are less portable and therefore exposed to exploitation (Pearson and Mitter 1993).

The self-employed, seemingly the most autonomous of the home-based workers, face similar issues to the rest of this population. Recent research has suggested a blurring of the distinctions between self-employed and employed: often these workers are subcontractors to larger producers and therefore subject to the conditions of independent contractors (Prügl and Tinker 1997). In addition, the women who constitute the vast majority of workers in this category are limited in their ability to be free agents in the marketplace, since they must adapt their income potential to the demands of their households.

Disaggregating the Telework Population

The telework trend is a variegated and complex set of phenomena. Rather than large segments of the population adopting telework, it appears to be suited to a specific population for whom the advantages outweigh the disadvantages. Not all the population will be in a position to work at home in the future because of space constraints and the nature of their occupations. Of those who could work at home, not all will choose to do so, because telework suits only some people. Even though working at home full time is not particularly advantageous to many people, working at home part time seems to be very popular, as flexible working patterns become more widespread.

Many aspects of telework have not been adequately studied. Though it appears that the availability of computer technologies is only one of the factors driving the shift to homework, alongside economic necessity and quality-of-life issues, the combination of factors is difficult to identify, because definitions vary and the homework population is difficult to establish. We know that it is advantageous for some segments of the population to work at home, but we don't know whether this decision is freely made or based on lack of alternatives. Though gender appears to be an important influence on the experience of working at home, we don't know how people adjust their images about home, family, work, and workplace, and whether these images are different for men and women.

We know that there are several distinct groups of teleworkers – employed teleworkers, independent contractors, and self-employed entrepreneurs – but we don't know the full extent of the differences between these groups in terms of work patterns. We know that feelings of isolation and overwork are common among teleworkers, but we don't know if this has to do with the nature of telework, the social structure and environment they are in, their psychological makeup, or a combination of factors. We know that teleworkers have mixed responses to the question of whether they have more or less control over their work, time, and environment, but we don't know why.

What seems to be missing in the analysis of telework is any comparison between working at home and in a corporate office. Though some studies have looked at the differences in working conditions between homeworkers and office workers (Gerson and Kraut 1988), nowhere has a comparison been made of how home-based workers and office workers use their workplaces, homes, and neighbourhoods. Naisbitt (1982) has argued that when people work at home they miss the social stimulation of the workplace. He further theorizes that home-based work weakens the symbolic ties that people have to where they work, including the use of workspaces to signify status, express self-identity, and reflect an organization's values. In this society, work has been explicitly defined as something done outside of the home to produce income, while unpaid domestic work is not clearly defined as work. The duality between work and home, and the male and female roles associated with these domains, is now being redefined as telework makes the home and the workplace an integral unit.

One study did find that women feel they must work outside the home in order to redefine sex roles, while male homeworkers maintained the same role at home as they had working outside the home (Gottlieb 1988). Nevertheless, because of the small sample of this study, more research is needed to corroborate this finding. Homework may be blurring, rather than reinforcing, traditional gender roles, especially for middle-class, self-employed professionals. In order to understand this issue an investigation of how male and female office and homeworkers perceive their home and work life is needed. Do people who work at home use neighbourhood resources more, and do teleworkers experience more spatial and temporal conflicts in their homes? A comparison between homeworkers and office workers would generate useful information on whether these conflicts are caused by the working situation or by other factors such as marital conflict over space.

Another unresearched aspect of home-based work is differences among household structures. Empirical research on environmental considerations has primarily addressed the issue of the home as a workplace and a place for family life for middle-class, dual-earner families. Those interviewed in prior studies also had a certain amount of flexibility, because of their income, to choose the kinds of spaces they wanted to live in. The following research identifies how the experience is different for single people, couples without children, and single parents. Teleworkers need to be studied in a range of economic circumstances, home settings, and neighbourhoods.

Methodological Approach
The research is framed in the context of qualitative policy research. Stack argues for public policies built around the social practices and everyday lives of people most affected by these policies and asserts that the ethnographic method "uncovers the complex dynamics of social change and the unintended consequences of social policies" (1997, 207). Using such an approach, this research consisted of qualitative research augmented by quantitative data that deconstructed the complex reality of telework. A better understanding of the effects of telework and home-based employment on

daily life and the use of the home and the neighbourhood has a host of policy implications from the macro to the micro scale.

Several studies form the basis for this book. An exploratory study of fifty-four respondents was completed in 1990 in the San Francisco and Sacramento, California, areas and comprised in-depth interviews, time/space diaries, and illustrative drawings to develop a profile of an individual and his or her environment. A mail-out survey with a response of 453, directed toward teleworkers and home-based entrepreneurs across Canada, was completed in 1995. A case study of the implications for telework for the Vancouver, British Columbia, area was done in 2000 and included an analysis of census data, a survey of current municipal practices, and in-depth interviews with a small sample of teleworkers (n = 8). Although these studies differ in time, space, jurisdiction, culture, and research approach, they have surprisingly similar findings. When the same questions were asked (for example about the number of hours worked per day), similar responses were received, which makes a strong argument for telework as a work situation that has distinct characteristics regardless of locale.

While these studies were not intended to be comparative, they complement each other in their use of quantitative and qualitative data and the resulting evidence is richer for it. The 1995 Canada-wide survey comprised a large sample to obtain statistically valid data. The 1990 exploratory survey and the 2000 case study research were designed as targeted research projects. Because of their small samples, the demographic data obtained cannot be generalized over the whole population. The Canadian survey covered the full range of home-based workers (including artisans, home-based business operators, etcetera) of whom 20 percent did not use a computer for their work, while both the exploratory study and the Vancouver case study concentrated on various categories of telework.

This research is about the fundamental dimensions of human life – space and time – and the transformations of these variables for teleworkers. The framework for examining the changing role of the home acknowledges the role of individuals in creating their world, but it also seeks to understand how the created social and physical environment shapes and selects forms of action in time and space. In this interchange between environment and action, human meaning takes shape and is expressed in an individual's felt sense of a situation. This interpretation provides descriptive insights into an individual's experiences and enables patterns to be generated from a range of experiences. The context, then, for studying home-based work lies within the specific social relations that determine an individual's activities in the home.

The data collected were analyzed to try to identify patterns, by thematically coding the material gathered. The themes that emerged from the material were used to develop profiles that described a range of situations reflecting distinct social, spatial, and temporal patterns in the home environment. Profiles are a way of organizing information in a conceptual structure that allows a range of interpretations on the issues. The profiles offer alternative possibilities for the variety of interactions between the individual and his or her environment.

In the course of the interviews for the studies in 1990 and 2000, I learned that people appreciate the opportunity to talk about their work situation. In many instances people collected articles for me or wrote down further thoughts that they later mailed to me. Especially in the 1990 study, many people were proud of being "homework pioneers" and eagerly and candidly talked about their experiences. Perhaps what is most significant about the homework phenomenon is that people's work is inextricably woven into the fabric of their lives. When I interviewed people, this plainly came out in their descriptions of how they lived. Sometimes these portraits were bleak: people stuck in a never-ending routine of work ... and more work. Other portraits were inspiring: people taking control of their situations so that they could lead balanced, creative lives. Many respondents had a hard time articulating their thoughts, and some of their most candid comments came at the end, after the tape recorder had been put away and I was packing up to leave. Though the interviews were structured, I found that when a particular aspect of their situation became significant I concentrated my questions in that area. By this, I learned (for example) about computer networks and their function as "electronic neighbourhoods." From other people I gathered ideas on the design implications of home-based work or thoughts on the changing nature of work organization when work is done at home.

The ten-year span encompassed by these studies reveals tremendous changes in how workers conduct their work. In 1990, computers were recognized as powerful tools, but their full power to convey information and communicate was neither fully understood by the general public nor implemented to any large degree. Now that work can be conducted entirely on-line, new relationships to work and the technologies that allow that work have been uncovered.

Results from all of the studies demonstrate that while the independent contractors and home-based entrepreneurs, on average, had a high level of education compared to the other categories, they reported the lowest incomes, the smallest housing units, and the least security in housing tenure. Female independent contractors and home-based entrepreneurs had significantly less income than the women from other categories and were much less likely to own their own homes. It appears that home-based work is a financially insecure phenomenon that affects the ability to provide security in other areas of life. In addition, the number of reported hours of work was consistently high in all three studies, especially for self-employed entrepreneurs, and there was a pattern of reported conflicts in trying to combine work and family responsibilities, especially for women. Rather than telework being a utopian ideal, it is very much a survival strategy for an insecure workforce. In the following chapters the experience of teleworkers will be profiled and analyzed based on the findings from these studies.

3
Working at Home and Being at Home: Blurred Boundaries

In the late 1980s, telework (then primarily called telecommuting) was in its infancy as a formal work option and many large corporations had started pilot telework programs to evaluate its effectiveness. Computers were in wide use but the World Wide Web had not been invented and Internet communication had not become mainstream. The socioeconomic imperatives of flexible labour were not yet as prominent as they became in the 1990s. However, while the data for this chapter were obtained ten years ago, the similar findings from the subsequent studies show that these data are still relevant to the current situation. The following profiles of home-based workers in northern California who used computers to conduct their work reflect a time when teleworkers had to define the parameters of their home/work interface in a work territory with very few guiding signposts.

Balancing Home and Work Life: Gendered Experiences

Nancy, thirty-five, starts her workday by walking down a few steps from the main part of her house in Walnut Creek, California, to her office, a converted family room.[1] Barely glancing at the serene view of her garden, she is totally engrossed in her work as a book editor for an educational publishing company. As a full-time salaried employee, she maintains a forty-hour-a-week schedule – if she is not interrupted by the demands of her two pre-school children. She has a babysitter but often her children bang on her door crying for attention. She describes her situation:

> If I didn't have kids I would just as soon be in an office but everything would pile up and I would not have time at all for my family ... Home and
> · work life [are] antithetical. Home life is chaotic, noisy and messy, but I have to have quiet to work. I have to separate the two. I can't succumb to temptation to work more.
>
> I've learned to put in my hours at my paid job, close the door, and then for the other hours do my other job of looking after my family, but I'm still

torn in all directions. I'm really two people at home. In my office, I am a professional. When I leave my office, I am a mother. The two roles are not compatible.

Nancy's decision to work at home evolved after her employer moved its head offices out of San Francisco. She worked for a time in a small office in the downtown area set up for employees who didn't move, but after her first pregnancy the tension between working and being a "mom" was too great. She decided to try working at home in 1984, five years before I interviewed her. Nancy has found that she cannot work at home and care for her children at the same time. She cites the primary determinant of her satisfaction with working at home as good child care. Half of her salary goes toward paying a babysitter.

Nancy, and many working mothers like her, recognize that they gave up important aspects of their lives when they opted to work at home. Their careers are now stalled. They miss the camaraderie of the office and the perks that made them feel like "adults," such as entertaining clients in restaurants and dressing in business clothes. Satisfaction in their work life is secondary, however, to the belief that working at home allows more time to be devoted to their families. Even Nancy, who has a structured workday, has found that not to be the case. The problems inherent in trying to balance a career and a family become even more apparent when they are in the same locale.

In contrast to this portrait, John, forty-five, had very different reasons for becoming a telecommuter. He describes himself as a workaholic. His decision to work at home was predicated on his long commutes and his desire to accept freelance commissions. His home is in Pacific Grove, a small seaside community a ninety-minute drive from the computer company in the Silicon Valley where he is employed as a technical writer/editor. He has arranged to be at the company headquarters for three days a week. During this time he works forty hours and stays with his mother, who lives close to his office. The rest of the time he works on freelance book assignments and lives with his wife and four-year-old daughter. Initially, he worked at home the rest of the week, but then found he couldn't get away from either his work or family demands for attention when he was at home. His solution was to rent a small office within walking distance of his home and occasionally work at home. John believes that his "satellite office" is a perfect compromise between working at home and working in a company office.

John loves both his work and his family but acknowledges that right now his work is coming first. His wife stayed home to take care of their child after she was born and has now begun to work, but only part time. John's involvement in child care consists of occasionally walking his daughter to her babysitter en route to his office. He rarely participates in housekeeping.

At present, he is working every day and on average seventy to eighty hours a week. He confides, "I really love my family, and I want to work less, but the money is good. My goal is to cut back to six days a week."

These portraits represent distinctly different attitudes toward work and family life. Nancy chose to work at home because, though she wanted and needed to work, her priority was the care of her family. John initially chose to work at home and then opted for an office near his home because he wanted autonomy in his work life. His working at home, and then close to home, did not change the relationship he had with his family. He would like to have more time for his family, but he sees his major role as the "bread-winner." Nancy is torn between fulfilling two roles, but her relationship with her family is also unchanged since she began to work at home. The tension between her work and family life has only become more visible.

Gender roles affect these perceptions. Female teleworkers see themselves as primary caregivers when they are at home. Men don't. Female teleworkers must cope with the dual responsibilities of work and family, while male teleworkers find it easier to separate these responsibilities as they see themselves primarily doing paid work at home. Gender differences affect choices in type of work and work location, and workers' ability to control those choices.

Not all male teleworkers, however, are unaffected by their experience. Some men choose to work at home because they want to be more involved in home life. Others find that once they start working at home they are more involved. Both Sylvia, forty-eight, and Tom, fifty-one, work at home in Davis, California. Sylvia works at home full time as an organizational consultant on home-based businesses. Tom works as a consultant and is also employed as the manager of a government telecommuting project that allows him to telecommute part time. They have converted the living room of their home into an office for their consulting business. Tom has converted a small bedroom into an office for his government work.

Sylvia has had a home-based business since their teenage daughter was born. Because she needed to take care of their daughter, she felt she had no other option. Tom started working at home two years ago, because he wanted to have control over his working environment and eliminate his commuting time. Now that he works at home, he has found that he is more involved in child-rearing and home chores but concedes that his daughter resents the time he spends working. His wife acknowledges that he is more accessible at home but still sees herself as the primary caregiver. Tom works sixty hours a week on paid work; Sylvia averages forty hours. Sylvia reiterates: "The more I work, the more things in the home slide. It is no different from people who have to commute ... but I am able to balance the needs of my family with work. I can do it because I am working at home. I can't do everything but I can choose what is most important and do what is needed."

Tom is primarily doing paid work when he is at home; Sylvia would like to be in that position, but she knows her primary responsibilities are elsewhere.

In contrast, Roger, fifty-three, concedes that he is a "househusband." Trained as an architect, he closed his office four years ago and began to write historical novels. Along with converting one of the bedrooms in his Mill Valley house to an office, he accepted the responsibility for maintaining the home. His wife has a full-time clerical job. He has learned to schedule his workday around household chores. He does a load of laundry before he starts work, cleans during his breaks, and quits work when he starts dinner. Though his children are grown, he has frequent visits from his daughter and grandchildren, which further limit his work time. He averages thirty-three hours a week on his writing and another fifteen hours on housework. Roger would love to write more, but recognizes that his home and family would suffer if he did.

While men and women with families have divergent gender roles, these roles become blurred for teleworkers without children. One such couple, Donna, forty-five, and Paul, fifty, are both technical writers. They met when Paul put an advertisement in the newspaper for a partner to share his work. Shortly after, they realized that they had many other things in common and became a couple. Both previously married with grown children, they now live and work together at home in Berkeley, California, because as they admit, "We want to work less, not more."

A growing number of couples are working at home together. In 1997, 40 percent of the total returns in the United States were joint proprietorship tax returns filed by husband and wives. About one-third of those were home-based businesses. Similarly, in Canada one in every three dual-earning couples (1.2 million households) is self-employed. Working at home gives both spouses an opportunity to be involved in a career and gives them more time together.

Donna and Paul's home and work life is blended. They work every day but rarely more than thirty-six hours a week. They start their day by discussing what needs to be done and dividing up the tasks. They share their work and household responsibilities and cook together. They work in the same office, which they converted from a bedroom in their two-bedroom apartment. They take frequent breaks, enjoying a coffee at their favourite café or wandering in their neighbourhood. Neither of them misses the office environment: Donna left a dead-end job for mental health reasons, and Paul wanted more independence in his work life. Now that they work at home they feel that they have more control over their work. They appreciate the ability to adjust their work to fit the needs of their personal life.

Donna and Paul have organized their daily schedule and environment in a way that meets their need for both a personal and a work life. Other couples have not been as successful. The pressures of a twenty-four-hour relationship can be taxing. Doug, fifty-four, and Marlene, fifty-one, both work at home in Sacramento but in different capacities. Doug works for a government agency and telecommutes part time. Marlene has a consulting firm that requires a lot of travel. Doug arranges his schedule so that when Marlene is at home he works at his government office. They have found that the tensions created in trying to use their telephone and computer at the same time become too great. Though they have a large, comfortable home, Doug acknowledges, "We tend to get in each other's way if we both work at home."

The experience of single people working at home is the opposite of too much companionship. Most feel isolated and have few social outlets. Sara, thirty-three, owns her own word-processing business in San Francisco. Doing almost all of the work herself, she averages sixteen-hour days, working six or seven days a week. For five years as a home-based entrepreneur she converted her living room into an office. She became a hermit during this period, rarely leaving her two-bedroom apartment, where she lives alone since a divorce. She got her food delivered and sent her finished work to her clients by modem or courier.

Sara became increasingly aware of how little socializing she was doing. Her friends weren't comfortable visiting in her "office." She was always available for her clients but she rarely had time for anything else. She began to resent the lack of boundaries between her home and business life. As a solution, she rented a small office in the downtown area, converted her dining room into a home office and restored her living room to a living space. She still works the same long hours both at home and at her office, but now finds that she is socializing more with her friends and in her neighbourhood.

While the experiences of the people described above are very different, they all have had the financial resources and flexibility in employment to allow them to solve some of their conflicts. Those who don't have a much harder time adapting to their situations. Ann, forty-three, a part-time teleworker, a single mother of two teenage sons, and a graduate student at the University of California, Berkeley, voices concern about the exploitive nature of working at home. She has a job with a research institute that requires her to analyze computer data and write reports. Because the institute is underfunded and lacks work space, her boss suggested that she work at home. She bought a computer with her own money. Now she brings home computer printouts and sends disks containing her reports to the office by messenger.

Ann lives in a tiny, two-bedroom, university-owned apartment in Albany, California. She works in her bedroom, which is just large enough for a single bed and a narrow desk. When she works she has to spread the computer printouts over her bed and on the floor. Officially she is supposed to be working only twenty-four hours a week. Often, she puts in many more hours at no extra pay. It is not uncommon for her to work twelve-hour stretches at the computer, stopping only when her headaches become too severe. While she enjoys the flexibility of her work schedule, at the same time she feels trapped in an "electronic sweatshop." As she describes, "My boss tries to make me feel like she is doing me a favour by letting me work at home. But I have to provide my own equipment and I don't get any benefits. When I am working at home it is usually a deadline and I have to work all of the time. Nothing gets done at home and I don't have time for my children. I have to work because I need the money but it is not a good situation." Ann did not choose to work at home, and because of her financial constraints she cannot afford to organize her home and family life in a way that would accommodate her situation. She feels trapped in a never-ending cycle of work, and resents the circumstances that force her to work at home.

As a skilled professional, Ann does have some flexibility in her work choices. Organized labour is opposed to home-based work because it is concerned about what will happen to workers who are less skilled. As companies and government agencies expand telecommuting programs beyond the ranks of mid-level managers and skilled professionals to include clerical personnel, these workers may be removed from official payrolls and designated as independent contractors without any benefits. They will be forced to pay for their own telecommunications equipment and its maintenance. Working on an hourly rate, they will be far less financially secure because of the lack of benefits.

While working at home is less secure financially, it is appreciated by segments of the population who would not be able to work except at home, such as the severely disabled. Two of the respondents in the study had disabilities. Don, a forty-year-old journalist and writer who is permanently confined to an iron lung, works in his living room in Berkeley where he also sleeps, eats, and entertains. He uses a wand that he manipulates from his mouth to tap out his stories on a computer. He gathers the materials he needs by telephone interviews. Once they are finished, he sends them by modem to a news agency he works for. He regularly uses the Internet and is a member of several user groups, through which he communicates with people who have similar interests. Though his circumstances are bleak and he makes only a subsistence income, his work has given his life a focus and meaning.

Retired people find that starting businesses from home gives them an opportunity to stay involved in work life. Frank, seventy-three, a retired professor, started a mail-order business for out-of-print university press books and finds that it has overtaken both his home and his life. He has converted two bedrooms into an office and stock room, and uses the basement to store books. He acknowledges that even though his wife is also involved in the business, she resents the time he spends at it.

Though the teleworker population is very disparate in terms of its resources, several distinct patterns appear to emerge. The prime motivating force is control. The ability to manage time and space according to one's own needs, and the feeling of control that affords, is the most important attribute of working at home. Teleworkers diverge, however, in terms of their priorities, and fall into two categories: those whose choices are primarily dictated by domestic concerns and those whose choices are dictated by work concerns. Only a few teleworkers have been able to balance the two. Those with domestic concerns, mostly women, choose to work at home because they believe that they can integrate family and work life. This choice is usually necessitated by economic need. Working at home is the alternative that allows them the most flexibility, but once they start working at home most find that home life disrupts the execution of their work. Work-centred individuals choose to work at home for flexibility and control over their work. The home environment is a facilitator of their work goals and reflects their professional identity. Most cannot cope with the office environment and corporate culture, and prefer the solitude of their home and the control they have over their home environment. These teleworkers find that their work life impinges on their home and personal life.

Financial resources affect the decisions of all categories of teleworkers. Those who can afford it organize their home life in a way that accommodates their work life. Those who can't must work in unsuitable environments with the constant pressure to perform both their home and work roles. The pressure to perform conflicting roles within the same environment exacerbates the feeling that their lives are "out of control."

The reasons that people choose to work at home have implications for the way they use their home both spatially and temporally. Family-oriented teleworkers have trouble defining themselves as workers. Their work settings are usually in spaces that are used for other activities, and they organize their schedules around the needs of their families. For work-centred individuals, work settings dominate their home environments, and their daily schedules are organized around their work.

Immersion in the Culture

The above profiles are based on interviews conducted between 1989 and 1990 with forty-five teleworkers and nine office workers in the San Francisco and Sacramento areas who used telecommunications and information technologies in the course of their work. These workers included full- and part-time workers, whose occupations varied from computer programmer to medical transcriptionist. The study had a sample that was 54 percent female and 46 percent male.[2] Both self-employed teleworkers and teleworkers employed by companies, in managerial capacities and as independent contractors, were interviewed. The research addressed two questions that pertain to the social and environmental impact of homework on everyday life: how is homework changing people's activity patterns, social networks, and living spaces; and what is the role of the home and neighbourhood in this context? The focus was on the differences between working in an office setting and working at home, doing similar kinds of work, and encompassed both social and environmental variables. (See Appendix A for the complete questionnaire, and Appendix C for a list of occupations included in the study.)

The methodology incorporated in-depth interviews, time/space diaries, and illustrative photographs and drawings to develop profiles of various segments of the teleworker population that describe the spatial and temporal patterns of their home, work, and social lives. Each interview lasted from one and a half to two and a half hours. The questions focused on the respondent's activity patterns, social networks, and living and working spaces. The respondents were also asked to complete a self-administered questionnaire on which they checked off adjectives that applied to them. The profiles that developed from this checklist have been included in this chapter in the section on psychological attributes.

The purpose of the interviews was to develop an understanding of how individuals adapt to home-based work and the use of information technologies in their homes. Because this project was conceived as an investigative study, it was not intended that the respondents could be generalized over the total population, but instead were chosen to represent various teleworker situations. Though the sample was not stratified, the respondents were selected based on occupation (i.e., information-based occupations), employment tenure (i.e., self-employed consultants and entrepreneurs, independent contractors, and employed telecommuters), household type (i.e., two-parent and one-parent families, couples, and solos), income (i.e., low to upper-middle income) and location (i.e., urban, mature suburban, suburban, and exurban neighbourhoods). Within this framework I interviewed an almost equal number of men and women in similar occupations. Employers were not interviewed. Though their perspective is important, my purpose was to develop an understanding of the teleworkers' experience. Further research needs to be done to compare this with the employers' experience.

The respondents were drawn from a wide geographical area within northern and central California including San Francisco, the East Bay, South Bay, Marin County, and the Sacramento metropolitan area. Several large telecommuting pilot projects of the California state government and Pacific Bell provided the sample for the telecommuters and the control group of office workers. Directors of these pilot projects arranged for contacts with the various departments who had telecommuting personnel. Organizations of the disabled and physically challenged provided the names of respondents who were disabled people working at home. Through personal contacts I obtained the names of self-employed entrepreneurs and retired people who had started businesses at home. From the initial respondents I then received more contacts, which led to clusters of workers in similar occupations.

Another source of contacts was two electronic listservs, one in the Apple Computer company in Cupertino, California, and the WELL Network, a community access computer network that serves a large number of teleworkers. A request posted on these networks provided several interesting respondents who were on the cutting-edge of the possibilities for the new information technologies. The most difficult people to contact were those performing low-paying piecework, such as medical transcriptionists (who transcribe doctors' reports). A serendipitous meeting led to a contact with an employee in the medical records department of a hospital, who gave me several names of home-based and hospital-based medical transcriptionists.

Thirty-one percent of the total sample (seventeen people) were telecommuters working part time at home, 43 percent were either self-employed consultants or home-based business operators, and 9 percent (five people) were independent contractors (Table 3.1). Two of the telecommuters were clerical staff. The independent contractors were all women employed to do medical transcribing and data entry. As independent contractors they worked for only one company on a piecework basis. In addition, in this study, nine office workers doing comparable work to their teleworker counterparts were interviewed.

The largest number of respondents were in the forty-five to fifty-four age bracket (42 percent of all the respondents – 40 percent of the men and 45 percent of the women), 30 percent were thirty to forty-four and 17 percent were fifty-five to sixty-four (Table 3.2). Four respondents were in their late twenties, and two were over sixty-five. As shown in Table 3.3 the general

Table 3.1

Employment status of respondents to California study, 1990

Employment status	Male		Female		Total	
	N	%	N	%	N	%
Employed teleworker	9	60.0	8	40.0	17	31.4
Independent contractor	–	0.0	5	100.0	5	9.3
Self-employed consultant	11	50.0	11	50.0	22	40.7
Home-based entrepreneur/ business operator	1	100.0	–	0.0	1	1.9
Office worker	4	44.4	5	55.6	9	16.7
Total	25	46.0	29	54.0	54	100.0

level of education for all the sample is high, with the majority having completed college or graduate school. Both men and women had comparable
levels of education, with men more likely to have attended graduate school
than women.

Table 3.4 illustrates that the majority of the sample lived in two-adult
households. Like the population as a whole, the lone parent households
were primarily female-headed. Forty percent (n = 22) had children under
twelve years of age living at home. Given a per capita personal income for
San Francisco County of $23,174 in 1987 and a per capita personal income
for the state of California of $18,753 in 1988, the incomes of the sample
were well above the average in California (State of California 1988). The
lowest reported incomes were from female independent contractors and
home-based entrepreneurs (see Table 3.5).

The occupations of the respondents can all be grouped under the service
and government sectors. They included technical writer/editor, management

Table 3.2

Age of respondents to California study, 1990

	Male		Female		Total	
Age	N	%	N	%	N	%
19-29	2	50.0	2	50.0	4	7.0
30-44	6	37.5	10	62.5	16	30.0
45-54	10	43.5	13	56.5	23	42.0
55-64	6	66.7	3	33.3	9	17.0
65 or older	1	50.0	1	50.0	2	4.0
Total	25	46.0	29	54.0	54	100.0

Table 3.3

Education status of respondents to California study, 1990

	Male		Female		Total	
Education	N	%	N	%	N	%
Grade school	1	50.0	1	50.0	2	3.7
High school	–	0.0	–	0.0	–	0.0
Some postsecondary	4	44.4	5	55.6	9	16.7
Technical school	–	0.0	2	100.0	2	3.7
College/university	2	22.2	7	77.8	9	16.7
Graduate school	18	58.1	13	41.9	31	57.4
Other	–	0.0	1	100.0	1	1.8
Total	25	46.0	29	54.0	54	100.0

Table 3.4

Household status of respondents to California study, 1990

	Male		Female		Total	
Household composition	N	%	N	%	N	%
Solo	7	41.2	10	58.8	17	31.5
Couple	10	66.7	5	33.3	15	27.7
Two-parent household	6	35.3	11	64.7	17	31.5
Lone parent household	1	25.0	3	75.0	4	7.4
Live with others	1	100.0	–	0.0	1	1.9
Total	25	46.0	29	54.0	54	100.0

Table 3.5

Annual household income of respondents to California study, 1990

	Male		Female		Total	
Annual household income, US$	N	%	N	%	N	%
Under $15,000	1	50.0	1	50.0	2	3.7
$15,001 to $29,000	1	25.0	3	75.0	4	7.4
$29,001 to $58,000	6	54.5	5	45.5	11	20.4
$58,001 to $85,000	10	40.0	15	60.0	25	46.3
Over $85,000	3	60.0	2	40.0	5	9.3
Undeclared	4	57.1	3	42.9	7	12.9
Total	25	46.0	29	54.0	54	100.0

and technical consultant, computer programmer, planner, architect, attorney, government analyst, word processor, and medical transcriptionist. The catch-all title "analyst" used by many of the professional government employees included a wide variety of skills and tasks; these workers develop, administer, and monitor government programs and regulations. Home-based workers often had several occupations, even if they were full-time salaried workers. In addition, these workers tended to give much more status to their job descriptions. For example, word processors called themselves data-entry consultants. This trend of elevating job descriptions was also seen in lower-skilled government employees who found that computerization had eliminated some menial tasks and added more complicated ones.

The respondents for this study had only recently started working at home when interviewed in 1989. At the time of the study just under half of the home-based workers had worked at home for more than five years, and twelve (29 percent) had worked at home for less than two years. Only a handful of those interviewed actively used the Internet but most had a fax modem.

Living a Twenty-Eight-Hour Day

Control over the use of time is one of the most obvious advantages of home-based work compared to the daily schedule of an office worker. However, most people find it hard to adjust to the freedom afforded by working at home. Generally, a home-based worker's day starts later than an office worker's because he or she doesn't have to commute. A teleworker can start anywhere from 7:00 a.m. to noon, work a few hours, take a break of several hours to do errands, and then go back to work in the evening. Some impose an office time frame on their workdays, working from 8:30 a.m. to 5:30 p.m. with an hour for lunch. Some people work better at night, and appreciate the flexibility to organize their workday in a way that reflects their body rhythms. A few live a "twenty-eight-hour day"; they get up, work, and go to bed when they want. Most, however, do not seem to be able to impose a schedule on their workdays and work long, irregular hours to meet deadlines. Some can't describe a typical day because their days vary so much.

Eric, thirty-two, a teleworker from San Jose employed as a computer programmer for a software company based in Sausalito, describes his routine in computer language as "scrolling through his day." He has no set time for any particular activity. Eric gets up around noon and spends about an hour and a half reading the newspaper and scanning his e-mail. He settles down to work at his terminal around 1:30 or 2:00 p.m. and works until he is hungry around 4:00 or 5:00 p.m., at which point he may have breakfast, lunch, or dinner. He may then watch the news for an hour and half or play video games (he has an extensive collection) with neighbourhood kids. He then works until his wife, who also works in the computer industry, comes home around 9:00 or 10:00 p.m., spends some time with her, and after she goes to bed is back at work until 4:00 a.m. He does no housework and his meals are either take-out or packaged food. He goes to his office about once every two weeks, but mainly communicates through his modem. His eating habits and lack of exercise are reflected in his physique: he is very overweight.

Another home-based worker, Sally, forty-nine, from the middle-class suburb of Hercules, California, is a medical transcriptionist. Her schedule is very different from Eric's because she has a daily deadline. In a typical day she gets up around 6:45 a.m., makes breakfast for her family, and does some household chores. She is at work by 9:00. She never takes breaks during the day, eating only a quick lunch. Every workday she must leave the house by 3:30 p.m. to deliver her finished work and pick up more tapes from the doctors' offices. She is home by 4:45, and prepares a quick dinner. Her husband helps her clean up and they talk. She starts work again at 7:00 p.m. and works until 10:00 most days. If there is a lot of work she will continue working until 11:00 or midnight. Immediately after she shuts down her computer she goes to bed. On Saturday she does errands and if there is a lot of work will work in the afternoon and evening. She always works on Sunday from noon to midnight.

Sally prefers working at home to being in an office because she likes working alone. Before she started working at home she worked the swing shift at a hospital, often getting home after midnight. She didn't see her family much. Now, at least, she says, "my body is here." Her husband likes that she works at home because he knows she is safe. Her children like it because they know she will be there when they get home.

Sally has organized her home life around her work. Since she doesn't have the time to do any major housework her mother comes over once a week to do the housekeeping and some cooking. Her husband does the vacuuming and the yard. Her daughter does errands such as going to the post office and the bank, and her son also helps out. All of them share the cooking with her. She admits that she has little time for socializing.

Ann, the single mother, graduate student and part-time teleworker profiled in the first section, has even less time to socialize. She works from 7:30 a.m. to past midnight analyzing computer data and writing reports, taking few breaks. The breaks she does take centre on her other role as a mother. Nevertheless, she rarely has dinner with her two teenage sons. Usually she only cooks a couple of times a week; her older son cooks for himself. When she is working, having her sons at home is a major problem. Their tiny apartment is too small to simultaneously accommodate her work and her children's activities. When she has a particularly pressing deadline she has to ask them to leave. Weekends are extremely trying because she has to work around the schedules of her sons.

In contrast to the enmeshing of work with home life in Ann's and Sally's schedules, Harry, twenty-eight, an electronic engineer, operates his business from his two-bedroom apartment in San Francisco on an office schedule. He gets up around 7:30 a.m., eats breakfast and is ready to start work by 8:30 or 9:00. He breaks for lunch around noon, going for a walk in his neighbourhood to a local restaurant. He comes back at 1:00 and works until 5:00 or 6:00 p.m. He never works in the evenings or on the weekend. Besides structuring his work time, Harry has structured his home environment to maintain a clear separation between his home and work. He keeps all work materials in the bedroom he has converted to his office. He closes the door to his office when he is not working and never goes into it after work hours. He has two phone lines, one for personal affairs and one for work. He never answers the personal line during the day and never answers the work line after work hours. He dresses to go into his home office differently from when he is just at home relaxing.

Few single teleworkers are as disciplined as Harry in maintaining a regular schedule. Brian, forty-nine, has a PhD in electrical engineering systems and has worked at home as a consultant since 1983. All he needs for work is a telephone and a computer. He is twice divorced and presently lives alone in a studio apartment in San Francisco that he describes as "an office which I live in," rather than a home where he works. Brian concedes that his work and personal schedules get confused. When he is at home he tends to be working all the time. He is not well disciplined in setting up a special time for work. His workday typically lasts from 7:00 a.m. to 5:00 or 6:00 p.m. He doesn't take breaks during the day, and if he is at home in the evenings he will work then as well.

Unlike teleworkers, office workers with long commutes generally have to get up between 5:30 and 6:00 a.m., and drive or take public transit for about one hour to arrive at work by 8:00 or 8:30 a.m. Often office workers don't take lunch because they find they can get their best work done when the office is quiet. They leave the office at 4:30 p.m. and arrive home by 5:30. They then have to unwind from the day's stresses for about an hour.

Karen, a thirty-eight-year-old lawyer with a thirteen-month-old baby, works in the office of a government agency in San Francisco and lives in Milpitas, a commute of three-quarters of an hour from her office. She describes a harried workday, in which she is up by 6:15 a.m., showers, and gets breakfast going. She gets the baby up, changed, and ready for daycare. She then gets dressed, makes her husband's lunch, and does some household chores. By 8:20 she is loading up the diaper bag and running out the door. She drops her husband off at BART (Bay Area Rapid Transit) and then drives to her office. Her son goes to a day care centre in the building where she works. She drops him off and is usually at work by 9:15. She doesn't take breaks; instead, she tries to see her baby during lunch. Karen is at the office until 5:30 p.m., gets home by 6:30, changes, and starts dinner. She watches the news at 7:00 and then eats. After dinner she and her husband clean up. The hour between 8:00 and 9:00 is devoted to the baby, including his bath. After he goes to sleep at 9:00 she may do some office work. Usually she does housework, watches television, or reads. She is in bed by 11:30. She wishes that she could devote more time to her work, but she doesn't have more time.

While most office workers feel constrained by their work schedule, part-time teleworkers appreciate the variety that their schedule affords them. Angela, forty-eight, a part-time teleworker, is an administrator doing management analysis and organizational development for a number of state government departments. She is a single parent with one daughter at home and a son in college. She lives in a single-family house in the quiet suburban neighbourhood of San Leandro. On the days she telecommutes from home she is very flexible. She likes to work at night and has chosen to participate in a telecommuting pilot project because she doesn't like the imposed time frame of a structured environment. She is just as likely to work on the weekend as on her two telecommute days. She works from twenty to sixty hours per week depending on her workload, and she may work anywhere from three to seven days a week.

On the days when she is at home, she gets up around 7:00 and starts work at 8:00 a.m. She takes a midday break to do errands and then goes back to work until 4:00 p.m. Usually she works after dinner from 7:00 to midnight. When she goes into her office in San Francisco she gets up at 5:30 a.m. to be on BART by 7:00 and arrive at her office by 8:00. She schedules meetings for the days she is in her office and usually works through the lunch hour or goes to an exercise class. She stops work at 4:30 and is at home by 5:15. Because of the stresses of the day it usually takes her an hour or two to unwind. She sometimes works at home from 7:00 to midnight.

Angela loves working at home and feels it is more convenient and a healthier way of working. She finds, however, that she has become more solitary since working at home. As she concedes, "I have a tendency to be a hermit and telecommuting accents that. I feel more withdrawn when working at home." Her daughter has also noticed that she has become withdrawn since she started telecommuting. Rather than wanting more socialization, Angela's vision is to move out to the country, telecommute for four days, and come to the office for one day.

Figure 3.1 outlines typical days for a male and female telecommuter and office worker, all with families and corporate-employed. While it is difficult to generalize a pattern for telecommuters, office workers have quite set routines. The male telecommuter illustrated works in one block of time from 8:00 a.m. to approximately 6:00 p.m. without taking a break for lunch. Instead, he takes frequent short breaks and snacks during the day. After quitting work he may help prepare dinner and do some chores. He may work at night, but it is more than likely he will do some kind of recreational activity, either at home or elsewhere. In contrast, the female telecommuter usually works at night to make up the time spent during the day doing household chores and errands. She starts work later, and while she does not take a break for lunch she stops her work for long periods of time during the day to handle household responsibilities.

The male and female office workers have similar daily work schedules, as they start and leave work at a set time. However, while the male office worker usually takes lunch, the female office worker uses that time to either work or exercise. In the evening, the male office worker, after eating and doing chores, may engage in some form of recreation while the female office worker must usually attend to household chores and the care of her children.

These profiles of home-based workers and office workers indicate many differences between the sexes and among economic classes. Of the men, Eric (the computer programmer) has the most flexibility, but his employment status has given him this control over his time. His home situation allows him to opt out of conventional time sequences, virtually living in a

Figure 3.1

Daily time/space patterns of typical telecommuters and office workers in California study, 1990

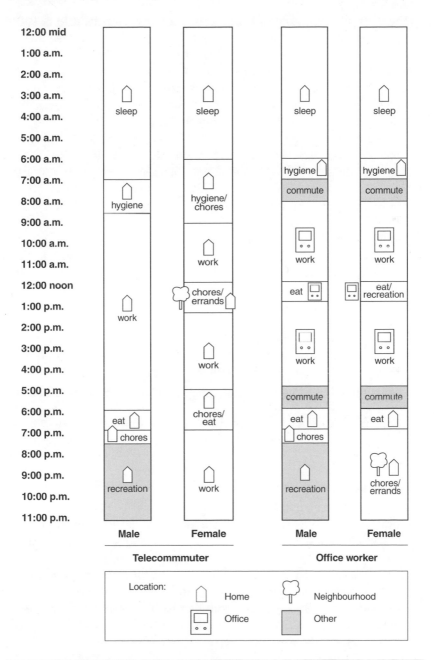

world of his own. Harry (the electronic engineer) also has the possibility for flexibility, but has opted to work a conventional forty-hour week. Brian (the electrical engineer) has allowed the demands of his work to control all other aspects of his life.

The women profiled did not have this flexibility. They tend to be more stretched in terms of the amount of time they have to do what is required of them. Sally (the independent contractor), Ann (the telecommuter), and Karen (the office worker) have rigid time demands placed on them because of the many responsibilities they have as both workers and homemakers. Ann, in particular, because of her lack of support and financial resources, has limited freedom to organize her time. Angela, the part-time telecommuter, is not as constrained as the others, partly because while working at home she also manages to structure her day to accomplish household tasks.

There are differences in the use of time between teleworkers and office workers. Only a few teleworkers maintain an office schedule like Harry's. Most have variations on Brian's or Sally's schedules. They work all day and most days and evenings, interrupting their work only when their personal or family responsibilities demand attention. Work appears to dominate teleworkers' lives, and though working at home gives them flexibility in how they manage their work time, they also can never get away from their work. In contrast, office workers have much more structured days and more separation between work and home activities. However, because of long commutes, they do not have more time for leisure activities, and most office workers also report problems in balancing their home and work responsibilities.

Findings from this research are corroborated by more recent studies. Wikström, Lindén, and Michelson (1998), in a study of teleworking in Sweden, examined the flexibility associated with teleworking. Their time-use survey data and the work diaries kept by homeworkers revealed that teleworkers felt they had to be available for colleagues and clients during work hours, and many found themselves working evenings and weekends as well.

Hours of Work

The profiles illustrate that most home-based workers work long hours with few breaks. Close to two-thirds of the telecommuters and home-based entrepreneurs work more than forty hours a week on average. Two-fifths of this sample report working more than fifty hours a week. The highest reported average was eighty hours a week. In contrast, only one-third of the office workers report working more than forty hours a week. While the large number of hours worked can be attributed to the self-employed status of many teleworkers, even corporate-employed telecommuters report working more than a forty-hour workweek.

While there are no appreciable differences in working hours between the sexes for both the full-time and part-time teleworker categories, there are differences for the office workers. Three out of the four male office workers report working more than forty hours per week, but none of the female office workers work over forty hours. This suggests that the hours spent working differs between the sexes when there is a structured work schedule, but is similar when there is more flexibility. Female teleworkers such as Sally, the medical transcriptionist, have the flexibility to organize their home life so that they can work more. Female office workers are forced to contain their work schedules by the logistics of trying to maintain a geographically dispersed home and work life.

Close to two-thirds of the telecommuters and home-based entrepreneurs work more than five days a week with over one-third of those working seven days a week. None of the office workers in the sample works more than five days a week. There is a definite pattern of teleworkers spreading out their work over the week, while office workers contain their workdays. Most teleworkers complain that the work is always there, visible to them. They can't leave it because, as one teleworker says, "It is staring at me all the time."

All of the respondents use a computer to do their work. In addition, 85 percent of the full-time male homeworkers, 60 percent of the part-time male homeworkers, and 30 percent of the full-time and part-time female homeworkers have a modem. In particular, the employed teleworkers communicate electronically to their office every day. They receive messages, memos, and documents from their corporate offices, and send back material they are working on. They recognize that communicating electronically somewhat alleviates the need for direct, immediate contact with associates and coworkers. They have found that e-mail has considerably changed how offices are run: much more information can be distributed quickly to more people.

Figure 3.2 illustrates the use of time and space over a one-week period for a typical male and female home-based entrepreneur/independent contractor, teleworker, and office worker. All of those diagrammed have children. The diagram was arrived at by averaging the number of hours spent in each activity for each segment of my respondent population. As can be shown from the diagram, the full-time home-based entrepreneurs/independent contractors consistently work the longest hours, stretching their workweeks into the weekend. Housework and leisure pursuits are de-emphasized by both the male and female home-based worker, but the woman does spend more time on housework and child-rearing. Spatial reach during the week is limited for both the men and women but the male teleworker has more flexibility in his movement due to the nature of his work and his freedom from household tasks and childrearing.

Figure 3.2

Weekly time/space patterns of typical home-based workers and office workers in California study, 1990

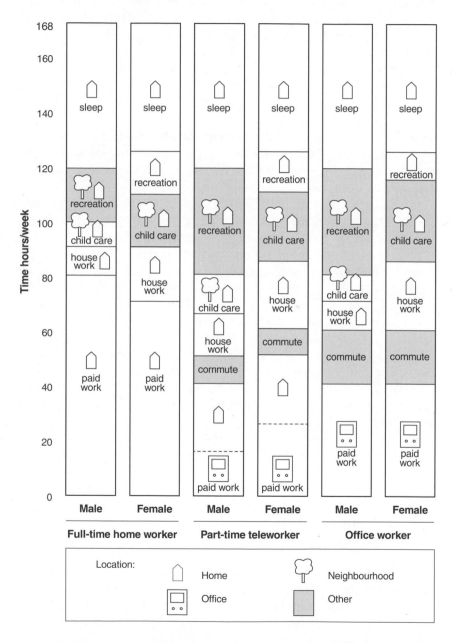

Source: Reformatted from Gurstein (1991).

The teleworkers work longer hours than their office counterparts when they are at home and also stretch out their workday with long breaks. When they go into their offices they spend a good portion of their time commuting, and unwinding when they return home. There is not as distinct a difference between the sexes for this group as for the others. The office workers consistently work fewer hours but spend a greater portion of their time commuting. The female office worker spends a larger portion of her day taking care of household and child care responsibilities than her male counterpart. While the spatial reach of the office worker is much wider, it is also limited to trips to and from the office and shopping. It appears that the time deleted from a person's day when he or she doesn't commute is used to work and not to handle the other responsibilities in his or her life.

Work Patterns

While almost all of the teleworkers say the freedom of managing their own time is a positive benefit of homework, many also talk about the fine line between flexibility and having work take over their lives. They are aware that they must learn to control the tendency to work all the time. Sara, the owner of a word processing business profiled at the beginning of the chapter, describes her situation: "Working at home allows me to work more. I have control of the amount of work I can do and when I can do it. But when I work at home most people don't realize that I live there as well. They think of it only as an office ... I am consumed with having my office at home. I must learn to say 'no' and modulate myself. I am used to working long hours and I am very good at what I do, but I am killing myself working too much."

Though teleworkers tend to work long hours, most feel that they are perceived by employers, clients, family, friends, and neighbours as not really working. They believe that their work is invisible to these people. Sylvia, the organizational consultant profiled previously, describes business meetings with clients in her home where her daughter, demanding attention, has insisted on sitting on her lap. Even Sally, the medical transcriptionist with the sixteen-hour-a-day, seven-day-a-week work schedule, sometimes finds that her family forgets that she is working, and thinks that she is just at home as a homemaker. When they want her to do errands for them she often hears the comment, "You have the time." Teleworkers are not recognized as real workers because they don't exhibit the role trappings of workers. Rebecca, forty-one, a busy self-employed technical writer and market researcher, concedes, "Friends don't respect that I am working at home. I am constantly being interrupted by personal phone calls and I am expected to handle personal emergencies. They would never call me during the day if I was working in an office."

Teleworkers who work part time at home, in contrast, find that they can concentrate more at home than in the office because there are fewer interruptions. They can turn off the telephone and work as long as they want to. The bonus for them is that they don't have the interruption of physically having to go home. When they work at home they can adjust their working patterns to what is most comfortable to them. Most feel better working in small time segments of two hours or less and taking several short breaks during the course of the day and evening. Most teleworkers, nevertheless, would not want to work at home full time because they would miss the socializing and feel out of touch with the corporate culture of the office.

However, what full-time teleworkers enjoy most about working at home is the ability to make decisions independent of others. Dale, fifty-six, a California government policy analyst and a full-time teleworker, states this succinctly, "I feel like I work for myself." He always had a claustrophobic feeling when he was in an office and resented having to be on a set schedule. He enjoys focusing on the work and feels that working at home gives him much more freedom to be creative. The office, to teleworkers like Dale, is not conducive to working because it contains too many distractions. They don't appreciate office socializing or the monitoring of work. They are intimidated by the office politics. Most, especially those who have worked in open-plan offices, find that the physical environment doesn't allow privacy. While most cite mental health reasons for deciding to leave the office setting, a few have physical reactions to the office environment, such as severe headaches caused by poor lighting and ventilation.

Home-based work is an escape from the hierarchical organization of the office environment and the managerial control imposed in that environment. Most corporate-employed teleworkers think that their managers are threatened by their working at home. Because managers have less control over what teleworkers do, they can't do their job, which is to manage. The majority of employed teleworkers view working at home as a perk for a dead-end job. They recognize that they have low visibility in terms of promotional possibilities, but they believe they are compensated by the freedom they have relative to their colleagues.

Most teleworkers find that they work very efficiently at home. There are distractions, but far fewer than in an office. Employed teleworkers find that guilt is a strong work motivator. They feel guilty about their pleasant work situation. They are aware that people in their office don't really believe that they are working when they are at home, and they feel they have to compensate for this perception by producing more.

Such an overwhelming consensus that working at home is more efficient and productive than working in an office corroborates several theories on work management. The "Hawthorne Effect" theorizes that when people feel they are in control they are more likely to be productive in their work (Adair

1984). "Theory Y" management assumes that people really want to do good work and do not need to be watched over, and that they should be evaluated solely on whether they accomplish the tasks their jobs require (McGregor 1985). Teleworkers want to demonstrate to their management that they are working.

Nevertheless, teleworkers find that they are limited in the kind of work that they can do. Their isolated location means that they work on segments without ever being able to see the whole product. Mary, forty-five, a data-processing consultant who works at home, recognizes that she doesn't have as much control over a project's process as she did when she worked in an office. She only sees small parts of the whole. Though she can do most of her work at home on her computer, the lack of contact with coworkers impedes her work because she cannot make decisions immediately. In addition, the kind of work that most teleworkers are doing is not people-oriented. Most feel isolated from the work world. They miss having people "to bounce ideas off of" and feeling part of a group effort. Teleworkers have freedom and control but they don't have a community. A few, however, don't feel the loss. When they were working in offices they tended to be overinvolved in office politics. Now, as Mary emphasizes, "Working at home has helped me be uninvolved in the workplace. It has allowed me to stand back and not be so caught up. It has given me a perspective on my work."

It appears that the most positive benefits offered by working at home are flexibility and control over work. For most home-based workers, however, their work setting, the home, is isolating and does not provide the needed social support. The most satisfied teleworkers appear to be those who only work part time at home and go to an office the rest of the time. While all of the home-based workers were committed to their situation at the time they were surveyed, a follow-up study in 1993 revealed that some have abandoned their home workplaces and taken employment that requires them to go to an office. Four of the eight people contacted are now working away from their homes. Their reasons had to do with financial insecurity, loneliness and isolation, and change of life situation.

Psychological Attributes
The respondents in the California study were asked to fill in an adjective checklist that was developed to describe an individual's attributes (Gough and Heilbrun 1983). The checklist offers words and ideas commonly used for description in everyday life in a systematic and standardized format. The 300 items in the list detect nuances as well as major distinguishing factors.

Several interesting patterns emerged when analyzing the data of the three main groups of respondents: employed teleworkers, independent contractors and home-based entrepreneurs, and office workers. On the modus operandi scales, which evaluate the respondent based on the total number of adjectives checked and the frequency of selection, high scorers are typified as adaptable, outgoing, reliable, and productive. Office workers and teleworkers generally had the highest scores. The lowest scorers on these scales were independent contractors and self-employed entrepreneurs. Low scores indicate that they are sceptical and fearful of the future, tend to evaluate themselves as deficient in social skills, and find it difficult to conform to the everyday expectations of interpersonal life. This pattern of low self-esteem, especially among independent contractors, repeats itself on the other scales. Employed teleworkers scored by far the highest in total number of adjectives checked, indicating that they are detail-oriented, with the ability to express nuances and shades of meaning. This finding correlates with the kind of work teleworkers do, since this group tends to be highly educated and in the most professional occupations.

In the need scales, a cluster of fifteen scales that address dispositions identified as important in need assessment, the pattern was opposite to that found in the modus operandi scales. While independent contractors and self-employed entrepreneurs may lack self-esteem, they score the highest on the achievement, dominance, intraception, exhibition, autonomy, aggression, and change scales. These are all scales correlated with self-reliance and autonomy. Employed teleworkers scored the highest on the endurance scale, which demonstrates a strong sense of duty and conscientiousness. Office workers scored the highest on the order, nurturance and affiliation, abasement, and deference scales. High scores on these scales illustrate a tendency toward rationality, need for order, cooperativeness, supportiveness, comfort in social situations, need to avoid conflict at all costs, and deference to others without loss of self-respect. All three categories of workers had the same low score on the succourance scale. A low score reflects independence and effectiveness in setting goals. The patterns reflected in these findings portray independent contractors as independent and self-reliant, teleworkers as very conscientious, and office workers as adept at interpersonal and social skills, and at dealing with organizational structures where cooperation and compromise is necessary.

Topical scales are a cluster of nine scales reflecting facets of personality and social dispositions. Of the five scales chosen to profile, employed teleworkers have the highest score for self-control and ideal self, reflecting an overcontrolled and diligent personality with the ability to attain goals. Independent contractors and self-employed entrepreneurs score the highest in self-confidence and creative personality. High scorers in these scales are initiators, confident, assertive, and venturesome. Office workers have the highest score on the personal adjustment scale. They have a positive attitude toward life, enjoy the company of others, and possess the ability to function well in both the domestic and work spheres. Office workers appear to be the best adjusted in terms of social situations. Teleworkers, independent contractors, and home-based entrepreneurs are the most driven to succeed. Independent contractors and home-based entrepreneurs are the most assertive in attaining their goals.

It appears that though independent contractors and self-employed entrepreneurs (who are predominantly female) are very career-oriented, they exhibit more tendencies to self-doubt and low self-esteem, and lack social skills compared to teleworkers and office workers. Their lack of self-confidence might be attributed to their gender and their comparatively low socioeconomic status. It might also reflect the fact that they spend so much of their time in isolation, and do not have an opportunity to interact with others. However, it might also be that people with this tendency choose to work at home because of their proclivity to low self-esteem, awkwardness in social situations, and need for autonomy in their work. Further study would be required to determine the causal relationship.

The Habits of Daily Life
While the portrait that has emerged of the typical home-based worker is of a work-oriented person who has little time for other activities, most homeworkers believe that working at home has improved the quality of their home life. The benefits are less stress and more convenience. All of the part-time teleworkers believe that working at home has improved their life; they are calmer and more rested. They feel less stress because they don't have to commute, have more autonomy, and feel more in control over their lives than when they worked in an office. Most, however, feel pressure to perform. While employed teleworkers have to produce to satisfy their bosses, self-employed entrepreneurs and independent contractors have to produce to survive. One-fifth of the self-employed entrepreneurs and independent contractors believe that working at home has made their life worse. Eighty percent of those are women. The reasons they cite are isolation, lack of a social life, and conflicts between home and work life. Most respondents from all three studies outlined in this book would recommend working at home to other people but feel that satisfaction with working at home depends on personality and the kind of work and home life a worker has.

Working at home is convenient. Household chores and errands can be done during work breaks. Meals can be prepared to the homeworker's liking. Times for exercise can be integrated into the day. However, only a few home-based workers take advantage of these opportunities. Most homeworkers have neither the time nor the inclination to do housework. The time homeworkers spend on housework is comparable to the time spent by office workers; those with small children spend the most time at housework – at least two and a half to three hours per day. All of the male and the single female home-based workers do less than one hour a day of housework. Females with families reported doing the most housework – at least two hours per day. They combine these activities with their paid work. For example, they do their laundry while working. A few homeworkers hire someone to clean their homes.

One of the reasons that women exhibit more conflicts in working at home is that they generally work more on domestic responsibilities. Research has found that employed mothers average forty hours each week at work, twenty hours on home chores, and twenty-five hours on child care (Robinson 1989). In contrast, employed fathers average forty-four hours each week on employment, eleven hours on home chores, and fourteen hours on child care. Women spend eighty-five hours a week on a combination of employment, home chores, and child care, while men's total is sixty-nine hours each week.

Only a few home-based workers take the time to prepare meals from scratch. They either skip lunch entirely or "zap" a packaged meal in the microwave and then sit down to work again. Often they go back to work after an equally quick dinner. Instead of proper meals, teleworkers snack many times during the day. Consequently, most have gained weight since working at home. They also rarely have a regularly scheduled exercise period during their day. A few have exercise equipment in their homes, while some take daily walks. The majority, however, work from early morning to late at night with few breaks. Women, especially, scheduled time for exercise when they worked in an office, either during lunch or after work. Now that they are working at home, they find it hard to schedule exercise time. Those who live in suburban locations have few recreational opportunities. Most find that their exercise time is the first thing they skip when they have to meet a deadline.

A sedentary life does have its toll. Irregular eating habits and lack of exercise, coupled with the pressure of having to produce, make most home-based workers prime candidates for some form of stress-related illness. A few have serious back problems. Some have regular migraine headaches. Interestingly, although working at home does not appear to be particularly healthy, almost all of the teleworkers in the study believe that homework is healthier than commuting to toxic office environments.

Another benefit cited by most home-based workers is the freedom to dress informally and comfortably. Most wear sweat clothes when they are working, or even pyjamas and housecoats. A few, however, find that dressing casually is bad for their self-esteem. Bob, thirty-five, a former full-time home-based worker and technical writer specializing in telecommunications, lives alone in San Francisco, and describes his situation: "When I used to work at home I used to dress in pyjamas until noon or later. Sometimes I wouldn't shower or shave all day. I looked a mess ... It got to the point where I didn't want to go out because I felt that I was a terrible dresser. I didn't feel that I fit in."

Self-esteem is a critical issue for teleworkers. They have few symbols of their professional identity. They don't have the visible signs of success in the working world, such as a large corner office and a secretary. They usually don't have a boss or a group of office compatriots who can praise and encourage their work. Teleworkers have to rely on themselves for motivation and feelings of self-worth. A few, like Harry, the electronic engineer profiled previously, maintain an office schedule and dress differently to go to their home offices than if they were just at home relaxing. These people have chosen to separate their professional identity from their personal life. Most teleworkers, however, are either unable or unwilling to separate their intertwined professional and personal identities. This inability to compartmentalize the different areas of life exacerbates their stress and problems with self-esteem.

Socializing and Recreation Patterns

Like the general population, few home-based workers socialize in their homes. Though they usually know their neighbours, they don't know them well enough to consider them friends. The friends they do see they usually see elsewhere. Most teleworkers don't feel that they have much time to socialize. Some don't invite people over because they are embarrassed about how messy their homes have become since they started working in them. Teleworkers who have converted their living rooms into offices relate how friends sometimes complain that they don't want to visit an office. Most teleworkers acknowledge that they don't have a large network of friends and say their network has become smaller since working at home.

Teleworkers have noticed that they have become more selective about choosing their friends. One confirmed teleworker describes his office friendships before he started working at home as "just a substitute for a community." Few of these office acquaintances survived his new lifestyle. Now, he might not have any more time for friendships than when he worked in an office, but he has found out what he feels is important in a friendship.

Single teleworkers feel the isolation of working at home the most, because they have neither the support of a family nor the camaraderie of the office. They resent that their work and solitary work environment leave them little opportunity to develop a social network. A few single teleworkers have developed special interests or become active in professional organizations where they meet like-minded people. Some have incorporated special activities into their day to relieve their isolation, such as taking their dogs to a dog park to meet other dog owners or going out to lunch in the neighbourhood. However, those who live where there are few amenities don't have these opportunities, and many feel "career-isolated."

Teleworkers with young children feel a different kind of isolation. They are much like homemakers, spending all day at home in the company of children. They miss the stimulation of adult company but can rarely schedule outings during the day. Those in suburban communities feel especially isolated. There is no one in their neighbourhoods during the day; most of the women are working outside of the home.

Office workers talk about friendships in entirely different ways than teleworkers. One female office worker describes how she only has friendships from the office but rarely sees any of these friends after work hours. Her socializing consists of frequent lunches with friends where she works and infrequent lunches with friends from where she previously worked. After work, she and her husband keep to themselves. Other office workers with families acknowledge that their lives are taken up with their work and their families, and they have little time for socializing. Single office workers have the most time but rarely entertain at home.

Neither the teleworkers nor the office workers studied socialize very much. When they do, teleworkers have developed social networks based on common interests that may have very little to do with their work. Office workers rely on office friendships to provide some of their socializing, but they also have developed social networks outside of their work. Similar to office workers, teleworkers with children rarely socialize.

Homeworkers generally typify themselves as "media junkies." They listen to the radio while they work, watch television news at night, and use the sitcoms and drama programs to unwind. The chief source of entertainment and relaxation in the home for most homeworkers (like most North Americans) is watching television. Homeworkers hooked up to the Internet spend much of their non-working time communicating with others on various e-mail networks lists and user groups. Since the introduction of the World Wide Web, they now surf Web pages and databases. Internet use in 1996 was estimated at 40 million users with growth of around 10 to 20 percent per month (Graham and Marvin 1996). It is now estimated at 163 million users (NUA Internet Surveys 1999).

Many self-employed home-based workers spend several hours a day communicating with others in these networks on a vast array of subjects from work-related issues to literature and personal problems. They talk about these networks as "virtual neighbourhoods," "virtual communities," and "virtual gathering places" (Rheingold 1993). George, thirty-six, a freelance radio producer and media worker from Oakland, California, describes his experience on the WELL Network: "The network is where I hang out and do my socializing. It is a gathering place for people who don't go out. It might as well be a physical place. I envision it as a village green. In cities, people hardly know the people next door. There really is very little need for that sort of community. Other communities and affinities, like this network, are much more important to me." The importance of computer networks as substitutes for the face-to-face interaction found in physically defined communities cannot be underestimated. Computer users see networks as a way to overcome the anonymity of modern life. They are on intimate terms with people living thousands of miles away whom they may never meet. The computer, to these users, is a vehicle that allows hundreds of people with like values and interests to come together in affinity groups.

The corollary to this phenomenon is that network users often prefer the type of interaction they have over the computer to face-to-face interactions. These virtual communities provide anonymity. People give themselves on-line pseudonyms and often relay fictitious scenarios about who they are and what they do. Whole romances can be played out in the safety of their homes without the risks involved in actually having to meet the romantic partner. Particularly lonely and isolated people use the networks as their sole outlet. Few people know who they are, except electronically. One California man who was a particularly active network participant shocked other users when he symbolically took his life by destroying the contributions he had made over the years to an electronic conversation run by a computer conferencing system (Markoff 1990). Several weeks later, he followed this "virtual" suicide by killing himself in the real world. Since then there have been numerous other incidents of erratic behaviour precipitated by extensive use of the Internet and World Wide Web (Turkle 1995).

These modes of communication do have consequences for the ability of home-based workers to conduct and participate in social interactions. Bob, the technical writer, believes that though he has become adept at manoeuvring through cyberspace and conducting all his relationships by the telephone, his face-to-face social skills have atrophied. He describes his life as a "telephone life," and feels lost in personal encounters. After working at home for more than four years he began to hate his circumstances. Under the pressure of tight deadlines, and unable to budget his time effectively, he found himself retreating to his apartment for two or three days at a time. He didn't have distinctive breaks and the loneliness became unbearable.

His solution was to find a job as a writer for a consulting company. However, he no longer functioned well in an office environment. He was intolerant of bureaucracy, office politics, and constant interruptions by coworkers. After ten months, he was fired. His supervisor told him that he was not a team player and would be happier on his own. Bob is now trying to work at home again. He has a new game plan where he is scheduling lunch appointments and joining professional organizations to keep his isolation at bay. He is apprehensive, however. He feels lost at sea, unable to fit into the old society, but unwilling to fully accept the new.

Bob reflects a growing segment of the population that has to learn how to cope with isolation and loneliness. The individual is increasingly being atomized both in work and domestic spheres, with expectations of autonomy and self-actualization. Coupled with this is the increasing physical isolation of the individual, as shared physical public and private space are replaced by telecommunicated interactions. The result is individuals with limited contacts in physical space. An article on the link between sex trade and information workers during the dot.com frenzy in San Francisco in the late 1990s documents an extreme form of social isolation (Rosen 2001). It asserts that many men in IT employment rarely have the time or social skills to develop intimate relationships. Instead, they rely on sex trade workers for sexual gratification. Sex in the new economy "is just another outsourced job perfect for independent contractors" (H8).

Assessment of Working at Home

In the late 1980s, the "frontier" attitude toward telework and home-based employment was prevalent among those who were interviewed. Part of that might have been due to the locale, in northern California, where the research was conducted. However, this perception also reflects the attributes of telework that allow new work relationships to form. Employed teleworkers see it as a perk that gives them autonomy from their offices. For independent contractors and self-employed workers, telework allows enough flexibility in their daily patterns that they can manage other aspects of their lives. As well, the ability to work remotely is a significant factor in maintaining work responsibilities in periods of crisis. For example, for a period of several weeks after the 1989 Loma Linda earthquake, the Bay Bridge, a major arterial bridge into San Francisco from the East Bay, was closed for repairs. Numerous newspaper stories recounted how people managed to work from their homes and other locales even when unable to go to their offices.

Table 3.6 summarizes the attributes of working at home. Respondents in the 1990 California study said the most important benefits of working at home were the lack of a commute and flexibility and control over work, the working environment, and use of time. Because of these factors they feel less stress. Other advantages to working at home include savings on clothes and other expenses and the convenience of being able to do errands during the day. Few homeworkers cited improved family ties as an advantage; many had unresolved conflicts between their responsibilities to family and to work.

The chief disadvantages to working at home are isolation and lack of opportunities for promotion in the corporate hierarchy. Teleworkers tend to feel more restricted psychologically. Working at home can be especially difficult for people with compulsive behaviours such as eating disorders, or drug or alcohol abuse problems. Also, homework does not provide the stimulation of interpersonal contact. Because there is no separation between personal and work life, career priorities tend to supersede personal ones, and they are more likely to overwork. Employed teleworkers in the California study also feel that colleagues respond negatively to their working at home and view them as not really working. Teleworkers compensate for the lack

Table 3.6

Attributes of working at home

Positive change	No change	Negative change
Flexible work patterns	Neighbouring patterns	Overwork
Control of time and space	Use of neighbourhood services	Isolation
Lack of stress due to commuting		Invisibility
Increased productivity		Lack of opportunities for promotion
Increase in new ways of socializing		Lack of office-type socializing
		Intrusion of work into home life
		Irregular eating habits
		Lack of regular exercise

of office socializing by finding new ways of socializing, such as computer networking. Teleworkers rarely socialize with their neighbours, nor do they use their neighbourhood services more than office workers.

The key issue is the conflict between home and work activities that is experienced by both home-based workers and office workers. For some, work life impinges on home life; for others, home life interferes with the execution of work. Female full-time homeworkers with families experience conflicts the most. They are constantly aware of neglecting home and family responsibilities when they are at work. Those who cope most successfully develop rigid spatial and temporal boundaries. Female office workers also experience these conflicts but are less conscious of them because their home responsibilities are not as visible. Male homeworkers and office workers are aware of their home responsibilities, but regard them as distractions from their work. Women tend to organize their workday more around the needs of their family than men, who see themselves primarily as working when at home. For men, the main advantage of working at home comes in terms of efficiency of work; for women, the advantage comes from being able to be more involved in home and children's activities.

These conflicts are not just internal to parents concerned that they are not spending enough time with their children. Most children resent the amount of time their parents work when they are at home. They complain that all their parents do is work. Young children demand attention by "acting out"; they insist on their parents' attention when they know their parents are working. Office workers don't have this problem as much as homeworkers, because when they work they are physically separated from their children.

Single home-based workers have different conflicts. Their work life allows them little time for a personal life. The isolation of their work location coupled with the blending of living and working is intolerable for some, who feel claustrophobic. They say they need more connections to the world and miss the social life in the office.

For others, the blending of home and work life gives a sense of involvement in both spheres. Donna and Paul, the couple who live and work together, appreciate how working at home allows them to maintain a rich home life and a creative work life. They accomplished this by making trade-offs in their professional advancement. However, recognizing that working at home does not provide financial security, in 1993 they were both corporate-employed and working in offices but with much less personal satisfaction than in their previous arrangement.

Most of the sample for this study would recommend working at home to other people. Most also expect that their present arrangement will continue for the foreseeable future. Part-time teleworkers would predominantly like to work either part time or flextime. All the full-time male homeworkers prefer to work at home full time, while only half of the full-time female homeworkers like this arrangement. In contrast, only half of the male office workers in the California study, but three-quarters of the female office workers, would like to work at home. Of those who would like to work at home, the majority would like it to be part time or flextime. As well, the majority would be interested in working in a satellite office close to their home.

When asked what they would tell someone planning to work at home, one of the first pieces of advice offered is to make sure that the work lends itself to working at home. The employer and the telecommuter should know and agree on what is expected. A spouse, if any, needs to understand that the homeworker is working at home and not available for other chores. If the homeworker has young children, child care arrangements are necessary, since it will be difficult to work and take care of children at the same time. Friends need to be told that the homeworker is working at home and can't be disturbed. Of critical importance is that home and work must be kept separate. A separate work area, with as few distractions as possible, should be set up away from the other activities of the home. Teleworkers need self-discipline. They need to learn when to cut themselves off from working and how not to waste time. They must learn to be self-motivators and to cope with being on their own. They have to learn their limits.

This research contradicts several common fallacies about working at home. One is the perception that those who work at home are not really working. Contrary to this opinion, teleworkers work long hours and rarely have time for other activities. Moreover, working at home does not allow more balance between the various facets of life. For some that may be the case, but for most, work takes precedence both temporally and spatially.

Teleworkers choose to work at home because they want flexibility and control over their work life. For those motivated primarily by domestic concerns, this means wanting to have a work life that allows an opportunity to maintain their family responsibilities. For those with work concerns, this means wanting more control over their work and daily schedule than when they worked in an office.

This chapter has interpreted the socioeconomic trends outlined in Chapter 2 through profiles of a range of teleworkers and their patterns of daily life. Tensions have been elucidated between desired flexibility and control over when and where work can be conducted and the problems of isolation, invisibility, and maintaining a balance between domestic and work responsibilities when work takes over every facet of life. There appears to be no easy relationship between home and work life. Work settings dominate the home environment in many instances, and teleworkers' daily schedules are organized around their work. Teleworkers work long, irregular hours, which usually means that they can never escape their work responsibilities. The pattern emerging for teleworkers is a way of life dominated by work.

The study conducted in northern California revealed a strong proclivity of teleworkers toward autonomy outside the confines of a corporate workplace. Their decision to be home-based workers can be seen as a form of resistance to the employment norm. Increasing labour flexibility in how, where, and under what terms work is conducted is precipitating a new social identity that constitutes a blurring of domestic and work life. While this blurring could potentially offer rich opportunities for integrating the public and private spheres, instead it is causing role conflicts and overlaps. In the following chapter these patterns will be further explored in the Canadian context.

4

A Strategy of a Dispensable Workforce: Telework in Canada

By the mid-1990s, the main impetus for new forms of work organization came from the restructuring that was occurring in corporate organizations and the reevaluation of values by established workers. North America had just had a period of significant "reengineering" of corporations, after which many people found themselves without formal employment. In Canada, a study found that in 1995 only half of primary earners in a household had a full-time job while 16.6 percent had part-time employment, in contrast to 3.8 percent part-time workers in 1953 (Duffy 1997). The people who were out of work and the corporations that needed to get work done after their substantial downsizing started to investigate other forms of employment. Often corporations opted for contract workers, a less committed but highly flexible workforce to whom they had no obligation to provide benefits.

Frequently "outsourced" workers also saw the benefits of contract employment in terms of flexible work hours and the substantial write-offs that self-employment provided. The rise in self-employment has been phenomenal. In the 1979-89 period, 17 percent of net job creation in Canada arose from self-employment, and 13.4 percent in the United States. Between 1989 and 1997, self-employment accounted for the majority of the net employment growth (79.4 percent) that took place in Canada but almost none (0.7 percent) of the net growth in the United States over the same period (Role of Self-Employment 1999). These dramatically divergent statistics can be attributed to the very strong economy sustained over the last decade in the United States, where there has been a large increase in job creation. The Canadian economy has not fared so well. Most of the new sole proprietorships created in Canada have started from home. However, it is predicted that one-quarter of new business ventures will fail in the first two years. Numerous articles in the popular press emphasize there is no guarantee of employment for life and put a positive spin on it with portrayals of workers who are on their fourth or fifth career.

Women's participation in the labour force has also affected self-employment. One-third of self-employed Canadians are women and women account for 40 percent of new business start-ups.

Another factor is the rise in early retirement. People in their fifties, many of whom benefited from escalating housing prices that made them financially secure, have started to opt for early retirement from their employment of many years and go into self-employment. This has created a highly skilled but dispensable workforce.

Along with insecurity in the workforce, workloads have substantially increased. Salaried workers, given that they are often doing their jobs plus the work of those in their companies who have been terminated, find they are spending long workdays at their jobs, which often spill over to the weekends. Those who are self-employed require long working hours in order to succeed. In Canada, a significant proportion of people is working fifty hours or more per week (Statistics Canada 1997). One in three Canadians identified themselves as workaholics, a substantial increase since the 1980s, and acknowledge more stress in their lives trying to balance their jobs and home life (Conference Board of Canada 1999). In the United States, a recent Harris poll (1999) found that the average American's work time has increased from forty hours a week in 1973 to fifty hours in 1997. In addition, given the rising cost of living, the average household has to have two incomes to survive. In North America, half the population did paid work in 1950; two-thirds work now.

Women have carried a disproportionate burden of the restructuring that has occurred both in the domestic sphere and in the workplace. The "feminization of labour" involves both an increase of women in the labour force and an increase in flexible employment practices such as part-time, seasonal, and casual work (Fudge 1991). Rather than women's increased participation eroding the sexual division of labour, it is consolidating it. Economic restructuring has also created more "women's work," such as sales and service sector jobs. This category employs 3.7 million workers, or 26 percent of the total employed – of whom two-thirds are women (Statistics Canada 1996). Given the precarious employment future, men are now going into sales and service, but rarely are women making inroads into traditionally male employment. Between 1991 and 1996 male participation increased in all of the lower-paid traditional female occupations (such as clerking, food and beverage, cashiers, child care workers, and secretaries) while women showed no signs of integrating into the highly skilled and well-paid jobs in the trades.

Corporations that recognized the loss of productivity due to the added stress and burnout of working conditions actively promoted part-time telework programs as an incentive to keep workers motivated. Following upon pilot projects developed in the United States and Europe (JALA Associates 1990;

European Foundation 1995), a number of pilot telework projects for employees of Canadian government agencies and private sector corporations were introduced in the early 1990s and subsequently evaluated. The studies found that the flexibility and freedom of telework enhances worker productivity. One study quantified this by as much as a 15 to 20 percent increase in productivity (Armstrong-Stassen, Solomon, and Templay 1998). The authors proposed telework as a response to downsizing, because "telework is an innovative and integrative way to preserve jobs while reducing organizational costs" (14). Though these telework pilot projects appeared to be highly successful, they were few in number and even fewer have survived as a long-term strategy. Managers felt uncomfortable with supervising remote workers, and the government agencies and corporations that were involved in these telework pilot projects have since further retrenched their organizations, laying off many more workers.

While information technology has led to a significant growth in employment for highly paid "knowledge workers," it is creating as many low-paying jobs in the sales and service sector, such as telephone service representatives. Most often they work in call centres, because of employers' need for close supervision and because the technologies (i.e., telephone linked with a computer) are there, but sometimes they work at home, mostly on a nonstandard temporary, part-time, and/or contractual basis. Data on the industry in Canada and elsewhere are unreliable, but it is currently one of the fastest growing in the country (Buchanan and Koch-Schulte 2000). The proliferation of toll-free numbers in North America certainly attests to the increasing use of these services. A 1998 Price Waterhouse Coopers study referred to in the Buchanan and Koch-Schulte report estimated that in Canada there are approximately 6,500 call centres, employing approximately 330,000 workers, many of whom are part-time (11). Close to half of those centres are in Ontario, 28 percent in the West, 19 percent in Quebec, and 5 percent in Atlantic Canada.

Call centres are differentiated by the type of service they provide and their relationship to the consumer. Inbound centres provide customer service and support (e.g., Internet and long-distance support and sales, financial and banking services, hotel and airline reservations, and dispatching) while outbound centres do sales and marketing (e.g., telemarketing, survey market research, and charity fundraising) (ibid.). Outbound centres are the most stressful for the workers because they require "cold calls," which are often met with resistance by the consumer, while inbound centres are contacted by the consumer. While the jobs require a high degree of skill in interpersonal communication and customer service, call centres are now being typified as the "sweatshops of the nineties" because they are low-paying, offer few opportunities for career enhancement, are closely monitored, and are highly stressful. These issues are little recognized when economic development plans are being formulated. The governments of some provinces, notably New Brunswick, are basing their industrial strategy on the promotion and retention of call centres.

The emergence of the World Wide Web in 1993 created a whole new set of relationships to technology. Even those who, in the 1980s, saw computerization as a threat – bringing job loss, increased automation of tasks, and decline in the quality of work – welcomed the Web's potential to make information accessible and create communities of like-minded people across a wide geographical area. Women especially were encouraged to embrace the new technologies as a way of levelling the playing field and extending their already well-developed abilities to communicate. In just a few years, however, the Web has come to be seen in terms of its potential to create and sustain economic growth. Gutstein (1999) maintains that the Internet primarily promotes a business agenda encouraged and sanctioned by government, specifically to develop electronic commerce, but also extending the market into schools, libraries, and research. The privatization of education and training institutions, libraries, and other social services agencies will be a direct consequence of this thrust.

Telework in Canada has to be looked at critically and in relation to other forces in society. It reflects the globalizing force of work practices as well as the localizing of those practices in daily life patterns. It is both a strategy by government and institutions to create a more "footloose" employment structure and a way for individuals to create more control over their work life. The following analysis of telework patterns uncovers a portrait of workers' employment and domestic strategies in that context.

The survey from which the following data are derived was part of the Canadian government strategy to understand the ramifications of telework.[1] It investigated the impact that telework and home-based employment have on the use of the home and neighbourhood, as well as gathering data on work characteristics. The study was designed as a mail-out survey to teleworkers and home-based entrepreneurs in all ten provinces. The survey instrument was predominantly multiple-choice and obtained detailed data on: (1) characteristics of the household; (2) a work profile of the home-based worker; (3) community context of household and housing type; (4) use of the home for work; (5) telecommunications usage for work; and (6) role of home and community life to the home-based worker. (See Appendix B for the survey instrument.)

The survey sample included teleworkers who work for public institutions, Crown corporations, and the private sector; independent contractors who work on contract to one company; self-employed consultants; and home-based business operators. Based on estimates of the various segments of the home-based work population, a representative sample of 1,677 was drawn from a wide variety of sources, including individual contacts with agencies and corporations across Canada, and twenty-six databases provided by regional economic development offices and home-based business associations. Potential respondents from these databases were randomly selected and directly sent a cover letter and questionnaire. The sample of independent contractors and teleworkers working for public and private corporations was derived from contacts made directly with these corporations. After a corporation agreed to participate, contact persons were sent cover letters and questionnaires, which they distributed to potential respondents.

The breakdown of the sample was based on an approximate equal ratio of male and female respondents, a concentration in urban areas, and weighting based on population concentration in Canadian regions. The largest portion of the sample was self-employed entrepreneurs (1,253 or 75 percent of the 1,677 questionnaires sent), the next largest was teleworkers working in either the public or private sector (374, 22 percent of the total), and the final portion was contract workers or independent contractors (50, 3 percent of the total). The Canada-wide response rate was 31 percent (n = 453; adjusted for wrong addresses and "not home-based worker"). Based on the response rate, the sample is reliable and representative within plus or minus 5 percent nineteen times out of twenty.

Of those who responded to the survey 55 percent are female. The relatively equal response from both sexes confirms previous studies that found that men and women tend to work at home at about the same rates, though women are more likely to work entirely at home (Deming 1994). There are definite patterns in work status according to sex. While 81 percent of the independent contractors are women working as low-paid semi-skilled contract workers, only 38 percent of the self-employed consultants are women. In the other categories the breakdown is much closer to half. Sixty-one percent of public sector teleworkers, 47 percent of private sector teleworkers, and 58 percent of home-based business operators are females. This latter figure is comparable to the Orser and Foster study (1992) and others, which document the significant increase in home-based women entrepreneurs. Statistics Canada (1997) found that more than 10 percent of professional and technical women were self-employed. Of women in the peak child rearing period (aged thirty-five to forty-four), 15.1 percent were self-employed.

Occupational Structure
As shown in Figure 4.1, the most cited occupations in the sample (23 percent of responses) are professional services such as educator, engineer, chartered accountant, architect, or lawyer; the next most cited occupations (23 percent) are business services such as computer consulting, word processing, and design services, followed by "other" home-based occupations such as communications, research, and client services (12 percent). Manufacturing/processing of crafts, food, and clothing makes up 11 percent, and retail

Figure 4.1

Occupation of home-based workers in Canadian survey, 1995

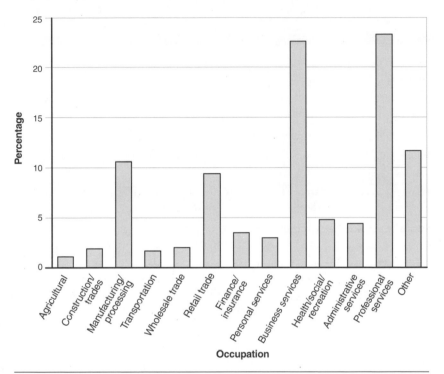

Source: Reformatted from Gurstein (1995).

trade/product sales 10 percent. The least cited occupations are agriculture, wholesale trade, and construction. Though this gives a comprehensive occupational profile, it should be recognized that there are limitations to any occupational classifications. For example, an engineer may be involved in both computer consulting and research, categories presented in the chart as mutually exclusive. Some examples of specific occupations included within the broader categories are listed in Table 4.1. The diversity of the occupations illustrates the wide range of activities that can be conducted at home. (See Appendix D for a complete list of respondent occupations.)

An at-home worker is more likely to have a professional or business occupation if he or she lives in British Columbia, Ontario, or Quebec. In the Prairies or Atlantic Canada, at-home workers are more likely to have occupations that involve manufacturing/processing of crafts and retail sales. On a regional basis, the breakdown of occupations differs only in Atlantic Canada. In all other regions the most cited occupations are business and

Table 4.1

List of occupations in Canadian survey, 1995

Occupation	Examples
Agriculture	Farming, nursery/greenhouse
Construction and trades	Contracting, plumbing
Manufacturing/ processing	Arts/crafts, clothing/dressmaking, food, health/beauty products, software producer
Wholesale trade	Wholesale selling
Retail trade/ product sales	Product sales, retail shop, sales representative
Finance, insurance, and real estate	Finance, insurance agent, insurance assessing, real estate development, real estate mortgagor;
Personal services	Cleaning, funeral director, gardening, introduction services, hairdressing, pet training, private investigator;
Business services	Bookkeeping, computing services and programming, conference planner, design services, financial management, graphic design services, word processing/ secretarial, training, marketing, video production/ photographer, writing/publishing
Health, social, and recreational services	Child care worker/foster parent, entertainer, health/ nutrition counselling, resort/lodge operator, social services consultant/equity consultant
Administrative	Government manager, government policy analyst, government research
Professional services	Architect/landscape architect/interior designer, chartered accountant, educational consultant, engineer, environmental consultant, lawyer

professional services, while in Atlantic Canada it is manufacturing/processing followed by business services.

Interestingly, on a gender basis the breakdown of occupations does not differ considerably: the most cited occupations are business and professional services and the least cited occupations are construction/trades and agriculture for women and agriculture and personal services for men. The similarity in occupations for male and female home-based workers might indicate less occupational differentiation when work is based at home, or, as mentioned earlier in this chapter, the fact that men are increasingly occupying traditionally female occupations.

Over one-tenth of the sample (13 percent) list more than one home-based occupation. Of those, 79 percent have two occupations and the rest list three to five occupations. Close to one-third of the sample (31 percent) have a home-based occupation that supplements their primary employment. This is comparable to the one-third of US home-based workers engaging in supplemental economic activities at home in 1991 (Deming 1994). The most cited of these primary occupations outside of the home are in the service sector, ranging from management to secretarial and waitress. Other primary occupations include work in governmental and educational institutions, and manufacturing.

The median number of years in the paid workforce is twenty, and the longest that a respondent has been in the paid workforce is fifty-two years. Only 4 percent (eighteen respondents) have been in the paid workforce for less than five years. The pattern of years in the paid workforce is similar for both teleworkers and other home-based workers. Home-based workers most likely have worked in their occupation for over ten years, but over half of the sample (57 percent) had another occupation or area of employment before their home-based work. The median number of years that the respondents have been in their occupations/areas of employment is seven. For most, working from home is a relatively new arrangement. The median for working at home in their occupations/areas of employment is 2.9 years, with two-thirds working at home in their occupations for less than five years. Teleworkers in the sample (i.e., employees in either the public or private sector) have generally worked longer in their occupations (on average twelve years) than the self-employed portion of the sample (on average six years). On average, both teleworkers and the self-employed have been working at home in their occupations for less than five years, but the self-employed tend to have worked at home for a longer duration than their teleworker counterparts. This reflects the relatively recent inception of most telework programs.

Of the 129 employed teleworkers in the survey, almost all are in a formal telework program. Besides formal programs, sixty-eight respondents had an informal arrangement to work at home with their employers. This includes informal arrangements that were made prior to formal programs, as well as informal arrangements that other home-based workers such as independent contractors have made with their employers. The median for formal programs was two years, and 2.6 years for the informal arrangements. Twenty percent of the teleworkers had been in a formal program for less than a year; another 28 percent had been in a formal program for less than two years.

Only 10 percent of the sample have home-based work of a seasonal nature (i.e., work at it only a few months of the year). Of those, two-fifths (39 percent) work for six months or less and 51 percent spend eight to ten months working at home with the median being eight months.

Figure 4.2

Work status of home-based workers in Canadian survey, 1995

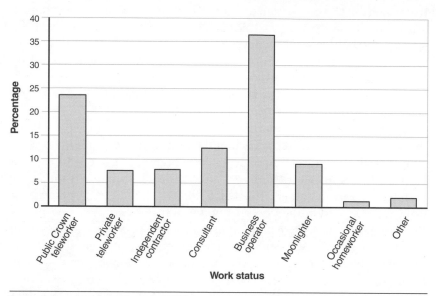

Source: Reformatted from Gurstein (1995).

Close to half of the respondents (48 percent) describe themselves as self-employed consultants or home-based business operators, and one-third (31 percent) describe themselves as public sector, Crown corporation, or private sector teleworkers (Figure 4.2). While the sampling frame did not include supplementers (i.e., employees who bring work home on an occasional basis), 13 percent of the respondents typify themselves as moonlighters, occasional homeworkers, or "other." Finally, 8 percent are independent contractors on contract to one company. These figures are comparable to the estimates developed in the Orser and Foster study (1992), in which 48 percent were self-employed, and the rest either supplementers or substituters such as teleworkers.

Demographics
In Canada the demographics of the home-based work population portray a mature, highly educated group with considerable experience in the paid labour force. Half are aged thirty to forty-four and close to one-third (31 percent) are aged forty-five to fifty-four (Figure 4.3). Their median age is forty-two, their minimum age is twenty-one, and their maximum age is seventy. This is similar to the California study, in which the largest number of respondents were in the forty-five to fifty-four year age bracket (42 per-

cent of all the respondents – 40 percent of the men and 45 percent of the women).

As shown in Figure 4.4, the general level of education for the sample is high, with the majority having completed college or graduate school.

Figure 4.3

Age of home-based workers in Canadian survey, 1995

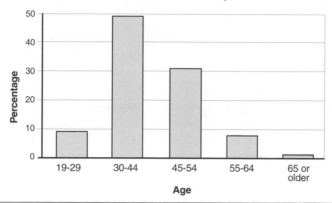

Source: Reformatted from Gurstein (1995).

Figure 4.4

Education level of home-based workers in Canadian survey, 1995, by work status

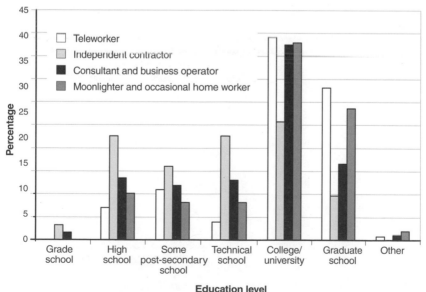

Source: Reformatted from Gurstein (1995).

Teleworkers have the most years of education, while independent contractors have the least. Both men and women in this sample attended graduate school in comparable numbers. This differs from the California study, in which men were more likely to have attended graduate school than women.

Figure 4.5 illustrates that the majority of the sample live in two-parent households. Like the population as a whole, the lone parent households are primarily female-headed. Fifty-five percent of the Canadian sample (n = 247) have children living at home under twelve years of age.

Like their counterparts in the California study, home-based workers in the Canadian sample tend to have an above-average household income; the majority derive less than half of their household income from home-based work (Figures 4.6 and 4.7). Teleworkers have the most household income on average, while independent contractors have the least. Only 14 percent of the respondents derive all of their annual household income from their home-based work. The largest percentage (63 percent) derive less than half of their annual household incomes from home-based work and of those, 16 percent derive less than 10 percent of their incomes from home-based work. That close to two-thirds of the sample derive under half of their annual household incomes from home-based work may reflect the preponderance of dual-income households in the sample, the large number of part-time workers, and the inclusion of home-based workers (such as teleworkers) who work a significant portion of their week away from home. The lowest reported incomes in both samples are from female independent contractors and home-based entrepreneurs.

Figure 4.5

Household status of home-based workers in Canadian survey, 1995

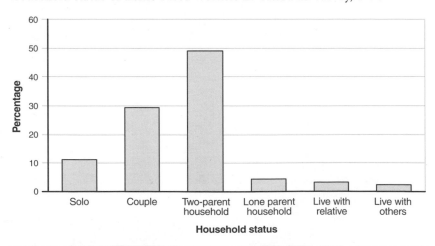

Source: Reformatted from Gurstein (1995).

Distinct differences emerge between employed teleworkers and other home-based workers. Employed teleworkers are better educated and have a larger annual household income. As well, they are generally in professional occupations, which allows them a fair degree of autonomy compared to the other categories of home-based workers.

Figure 4.6

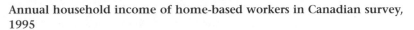

Annual household income of home-based workers in Canadian survey, 1995

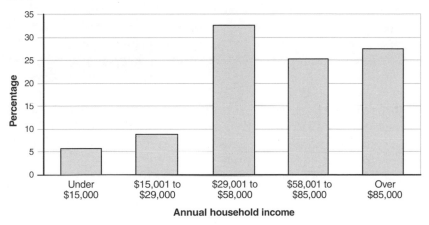

Source: Reformatted from Gurstein (1995).

Figure 4.7

Percentage of household income from home-based work in Canadian survey, 1995

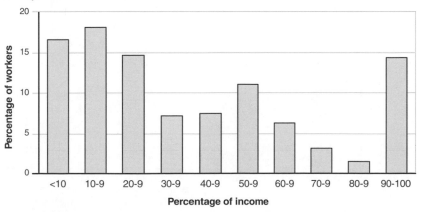

Source: Reformatted from Gurstein (1995).

Work Profile

The findings demonstrate a significant number of hours worked per week above forty. As Figure 4.8 illustrates, self-employed consultants and business operators, and moonlighters and occasional homeworkers, average the most total working hours per week (forty-six hours and forty-five hours per work respectively). Independent contractors and employed teleworkers work the least (thirty-three hours and thirty-seven hours per week respectively). Female and male home-based workers work on average similar hours (44.46 hours and 43.75 hours per week respectively), but women are more likely to work all of their time at home than men. The relatively small number of hours worked by independent contractors reflects the part-time status of most of these workers, working on piecework or "on call." The median total number of hours per week worked by respondents is forty-four, with the highest reported figure ninety hours per week. The large percentage of home-based workers in this sample who work over forty hours reflects a flexible workforce whose work schedule is largely dictated by the demands of their work, rather than predetermined work schedules. Generally, hours of work are increasing for all workers in North America, and home-based workers, especially those who are self-employed, are part of that trend. One-quarter of the respondents in a recent study said that they worked more at home without any decline in time spent at the office (Internet 2000).

Figure 4.8

Average total work hours per week at home and at work, by work status, in Canadian survey, 1995

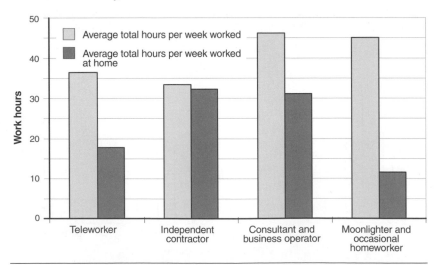

Source: Reformatted from Gurstein (1995).

The number of hours worked from home varies according to the work status of teleworkers. Figure 4.8 illustrates that independent contractors work the most number of hours at home; employed teleworkers who have an office in their corporate headquarters and homeworkers who bring work home on an occasional basis work the least. Sixty percent of the respondents work at another work location besides their home, with the median number of hours per week worked at another location twenty-four hours. Over half (53 percent) of those who work at another work location work there for less than thirty hours per week. Of those, close to one-fifth (22 percent) work more than forty hours per week overall – i.e., at home and away from home. The work patterns identified in this sample portray a very mobile workforce often working in several locations.

The emergence and affordability of new technologies is perceived by many writers as a major factor encouraging home-based employment (Herman 1993). The survey results confirm that home-based workers are heavy users of these technologies, though the reasons this sample started working at home pertain to control over time and space, not the availability of technology. Almost all of the home-based workers (84 percent) have a computer to conduct their home-based work activities, 77 percent have a telephone answering machine or voice mail, 31 percent have a fax machine, and 31 percent have a fax modem. A sizable number (28 percent), mostly teleworkers, regularly use electronic mail and the Internet.

The large percentage that use a computer, regardless of the type of work they do, reflects its importance as a tool for home-based workers. The use of the fax machine and fax modem for conducting work is also important, but their use decreased somewhat in the late 1990s as sending attachments via e-mail became common practice. As well, over one-quarter of the survey population uses e-mail or computer networks to communicate with colleagues, associates, and clients. Teleworkers and self-employed homeworkers use similar equipment to conduct work. However, more teleworkers use computers and more self-employed homeworkers use fax machines, photocopiers, and typewriters in their homes. Though a large number of respondents in this study use computers and personal communication devices, if the study were repeated today the number of users would probably be much higher, as the use of the Internet and cellular phones has increased significantly throughout the world.

The different patterns of use in the 1990 and 1995 surveys reflect the rapid growth of the Internet and the World Wide Web over those years. In the late 1980s, when the California study was conducted, it was still mainly the territory of academics and computer experts. Now, with the recognition of the Internet's potential for a wide variety of information dissemination, including for-profit opportunities, its use for communication, networking, and research has grown phenomenally.

Not all homeworkers appreciate the capabilities of the new technologies. One homeworker with a fax machine talked about how his employer, a night person, faxes him memos at all times of the day and night. Because he never knows whether a message is important, he feels he has to respond to all of them. Effectively he is on the job all of the time. Self-employed entrepreneurs feel similar pressures; clients expect them to respond to faxes and e-mail messages immediately and question their commitment to their projects if they don't. For the home-based workers in this sample, work takes up a significant portion of waking hours, both at home and other working locations.

Work Organization
Respondents spend on average over half of their time (53 percent) in their home offices or workshops, 11 percent travelling to job sites or out of town, and one-third (34 percent) in other work-related places. Teleworkers generally work part time at home, going to corporate offices for meetings, and to confer with superiors and colleagues. Self-employed consultants and home-based business operators travel a considerable amount for their work, visiting clients and associates. Independent contractors work almost exclusively at home, travelling only to deliver completed work and obtain new projects. They spend the most time in their home offices, followed by home-based business operators and public sector teleworkers. Private sector teleworkers, occasional homeworkers, and moonlighters work in their home offices the least number of hours.

Employing others at home can be an important consideration in the design and enjoyment of one's home and is a consideration in zoning regulations regarding home-based work. Sixty-five consultants and home-based business operators (14 percent of the total sample) employ others in their home workspace. Over three-fifths (62 percent) of those employ one employee either part or full time. The remainder have two or more employees, with the largest staff being seven full-time employees and six part-time employees.

Employed teleworkers' experiences are slightly different from those of other home-based workers. These teleworkers generally find that their productivity has increased compared to office work, because they have fewer interruptions, less stress, and a better environment for concentration. One teleworker in the Canadian survey commented, "I have completed tasks which have required great concentration such as reading and writing in less time and with better final results." Others find it hard to measure their productivity. One teleworker wrote, "I lose time in converting texts before leaving and returning to the office, and also in sending texts through modem," and another pointed out, "It's only one or two days per week, and given type of work productivity is hard to measure."

Because most teleworkers are actually only working at home one or two days per week, most do not feel isolated or miss the social interaction in their corporate offices. Also, because they only work part time at home, they do not believe that working at home reduces their managers' awareness of them and their work. However, some are concerned that meetings that they should be attending are scheduled on the days when they work at home, and that they are less visible in the office. As well, many (59 percent) find that they are working an average of three extra hours per week on those days that they work at home. This overtime is almost always initiated by themselves.

While both teleworkers and self-employed home-based workers value the opportunity that working at home provides for flexibility in their work life, telecommuters who live far from their work particularly recognize the advantages of working at home for several days a week. However, some teleworkers are concerned about the strict organization of the telework program that they are in and the lack of understanding by their supervisors regarding their status when they are working at home. For example, a daily newspaper in Vancouver, British Columbia, instituted a telework program for its reporters and other personnel, but ended the program after only a few months because supervisors became concerned that they could not reach their employees during the day. When they called and got answering machines, they immediately assumed that their staff were not working. Evidence of productivity when at home was not enough to convince the corporation's executives of their staff's effective use of time.

Few of the teleworkers (15 percent) have noticed any significant changes in their work group (such as effects on teamwork and communications) due to working at home. Those who have noticed a difference have found that they have to be more organized. As one homeworker commented, "I have to be more organized about scheduling meetings with my employees. I have had to think ahead about what tasks could best be handled on my work at home day given lack of access to [the computer program] Coronet and folders." Another teleworker has found, "Some office staff feel that my work at home day is a day off. Not so! I get much more done at home. I feel there is some jealousy – animosity from other employees."

Those who have not noticed any changes in their work group have found that they can easily communicate with colleagues and clients when at home. One teleworker commented, "My work colleagues can contact me by phone if they want my help, and often clients do not realize that I am at home because my calls are transferred directly to my residence." Another teleworker agrees, "Much of what I do, I do alone, after consultation with others. I use my days at work to network. Things will improve significantly when I get a dial-in [modem]."

Two-thirds of the teleworkers have a computer at home, and another 20 percent need to have a computer but don't currently. Just under two-thirds (63 percent) already have an answering machine at home; another 9 percent need to have one for their work and 16 percent would like to have one. Over one-quarter (27 percent) already have a modem at home; another 30 percent need to have one for their work. One-fifth already have a fax machine at home; 17 percent need to have one for their work and 60 percent would like one. Only 7 percent have access to their organization's mainframe at home; 39 percent need to have access for their work and another 39 percent would like to have access.

Employed teleworkers in general are very satisfied with the opportunity to work at home and believe that their productivity has increased. They also seem to be satisfied that working at home will not have undue consequences for their career opportunities and interactions with managers and coworkers.

Household Activities

The patterns of household activities, such as chores and child care, influence and help explain how different segments of the telework population use their home and community. As Figure 4.9 illustrates, the female respondents spend more time on household chores and child care than the men. Home-based workers in this sample spend, as a median, eleven hours per week doing household maintenance, and those who have child care responsibilities (55 percent of the total, n = 247) devote nineteen hours per week to them. While over two-fifths (43 percent) of the respondents spend

Figure 4.9

Average total hours per week engaged in household maintenance and child care in Canadian survey, 1995

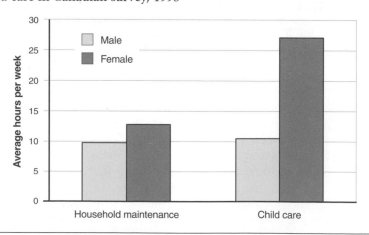

Source: Reformatted from Gurstein (1995).

less than ten hours per week doing household chores, the rest range from ten to forty hours per week of chores.

Child care responsibilities affect the work patterns of home-based workers, resulting in potential conflicts between work and domestic life. One-fifth (19 percent) of the respondents have children at home who require care. Close to half (46 percent) of those handle child care themselves or have a spouse (usually female) do it. For the rest, arrangements include child care centres and after-school care, babysitters, relatives, neighbours, or nannies. When workers begin to telework many believe that it will allow them more time to spend with their families. One female independent contractor, however, makes it clear: "You cannot mix child care and work. Either you are working at home or you are taking care of kids. You cannot do both."

An average (and median) of twelve hours per week is spent on leisure activities and volunteer work. Men spend more time on leisure activities than women. Over two-fifths (42 percent) of the respondents spend less than ten hours per week on leisure activities and volunteer work; the rest range from ten to forty hours per week.

Flexibility and Control

Results of the survey reveal the nature and extent of home-based work activity (including telework) in Canada and indicate a high level of satisfaction with working at home among the teleworkers and home-based entrepreneurs surveyed. Their concerns are their inability to disconnect work from home activities and a sense of isolation. This corroborates many of the findings in the qualitative exploratory study described in Chapter 3, but generally portrays a more positive perspective, which could be attributed to the nature of this survey and the self-selection process of the respondents. Since the Canadian survey was a mail-out questionnaire of mainly multiple-choice questions, there was little opportunity to probe for more complex answers. The questionnaire, though, did provide space for comments that reveal the attitudes of the respondents. Also, as this survey was based on a random sample of teleworkers and home-based entrepreneurs, it is expected that the respondents self-selected to a certain extent. Those who had particularly negative experiences with home-based work might not be as willing to answer a survey on their experiences as those whose experiences are somewhat positive. As well, the 1995 Canadian survey might have received more positive responses than the 1990 California study because home-based work is now generally more accepted. It is no longer an oddity, it is frequently mentioned in the popular press as the wave of the future, and it is portrayed as a highly desirable option. This might help people who are working at home see themselves in a positive light.

As in the California study, the Canadian sample finds that working at home provides flexibility and control over work, allowing more integration between home and work life. Female home-based workers with families particularly appreciate the opportunity that home-based work affords to maintain their family responsibilities, even though many acknowledge that this can be problematic. Male home-based workers cherish the control they have over their work and daily schedule. Teleworkers especially recognize reduced stress and fewer interruptions, resulting in better productivity, when they work at home several days a week. Teleworkers who live far from their work see the advantages in reduced travel. The disadvantages of home-based work are maintaining a healthy balance between home and work life, and the lack of social interactions. In general, the advantages considerably outweigh the disadvantages for most home-based workers.

Unlike the California study, in which many of the respondents voiced ambivalence about their at-home situation, satisfaction with working at home is very high for the sample in the Canadian-wide telework and home-based employment survey. Four-fifths of the respondents are very satisfied with working at home, and the same percentage see working at home as a long-term employment strategy. Comments on their situation include:

> My stress level has reduced a lot since working from home. It's great! It offers the best of both worlds.

> Working at home gives you the flexibility to set your own hours, and schedule around family activities.

Those who are not satisfied with working at home sound a more cautionary note:

> I love my work but I need to disconnect from it, which is nearly impossible.

> Too much home!

> Working at home very much reduces the pleasure of work. Human contact is important.

> Work too many hours and isolated.

The most cited reasons for starting to work at home pertain to control over time and space. These reasons include: convenience (52 percent of the respondents), flexible hours (46 percent), control over work (45 percent), cost considerations (35 percent), and control over work environment (34 percent). Other reasons include "Unable to find salaried position which offered the scope and challenge looking for," and "Wanted to work while raising children."

Typically, the reasons cited reflect gendered priorities. Female respondents particularly enjoy that they can be available for their children, and schedule their own time. One female respondent commented, "I don't have to arrange for child care, flexible hours, don't have to travel, don't have to dress up," and another woman echoed these words with, "I can schedule my own time and be home with my family – it's more flexible, less stressful, offers more independence and less waste of time." A woman with a disability said, "I have no other options, it allows me flexibility and the opportunity to work from my bed." Another woman commented, "Working from home reduces the necessity of breathing recycled air-conditioned air and auto exhausts. It's wonderful!"

The male home-based workers see the positive benefits but their comments focus more on convenience and reduced costs. One male respondent wrote, "Working from home offers convenience, flexible working hours, reduced transportation time and costs, and access to kitchen for refreshments, snacks, etc." Another listed, "Don't have to commute. Relaxed pace. Can cycle to park for lunch breaks," and a third is enthusiastic about: "Better quality of life. Costs less. Saves money on clothes." Some men who are working from home see the positive benefits of being close to family activities. One comments, "I can watch my children grow, and I am more focused on the needs of both family and business," and another man agrees, "I can combine child care with most work activities."

The disadvantages of working at home include the difficulty of separating one's professional life from one's personal life and the lack of social interactions. For the self-employed entrepreneurs, there is the added stress of financial insecurity. Isolation is also a major concern, as are lack of professional image when at home, reflected in cramped working spaces, and lack of self-discipline and motivation precipitated by being at home and away from colleagues and business associates. One female worker commented, "Work never ends – late evenings, weekends, no office hours for clients to call," and another agrees that the disadvantages are "the distractions – difficult to detach from work, phone calls at all hours, weekends, etc." One woman wrote that she "sometimes misses the interaction with coworkers," while another agrees, "Not enough socializing. I feel that I have to work when not doing anything else." A significant comment made by one female worker is, "I don't have any leisure time at home. I have to go away from home to get a treat."

The male homeworkers echo many of the comments made by women. They too are concerned with "allocating time between working and nonworking hours – separating work from leisure." They have difficulty differentiating between personal and business life as they have "customers who will show up at all odd hours of the day, seven days a week," and feel as if they are "always at work – must learn to put time aside for family and self." They are also concerned about family interruptions, increased noise and disturbances, and more distractions resulting in, "Less pressure to produce, less sense of belonging."

Problems encountered working at home are addressed by a variety of strategies and modifications. Better time organization is perceived as one solution. Some of the social problems, such as frequent interruptions and overwork, are difficult to resolve and require better communication and ability to set limits. Female home-based workers in particular find that they have to "train" their families to respect their work priorities more, and they have learned to voice concerns to supervisors over such inconveniences as scheduling meetings on days that they work at home. They have to schedule their time and resources in a more efficient manner, such as establishing office hours and telling their families when they are working at home. This, however, does not always work out satisfactorily. As one female home-based worker commented, "I try to schedule work activities, but I do work early in the morning or late in the evening."

Despite the high degree of satisfaction with working at home, when the sample was asked to assess a number of specific factors in their work, home, and community life, the response was mixed. Three-quarters of the respondents find that their work life is less stressful and two-thirds notice that they have more time to devote to household responsibilities and personal business when they work from home. Over two-thirds of the sample (67 percent) believe that other members of their household and friends have learned a great deal about their work, and that the lives of household members are less stressful when they work at home.

Half of the respondents, however, find that household responsibilities and chores often distract them from work activities, and that other members of the household and friends frequently interrupt their work activities. In addition, close to two-thirds (62 percent) believe that they work too much when they work at home and that their time with members of their household is often interrupted by work activities. Just under half of the respondents (45 percent) often feel lonely when working at home. One-third feel they have fewer friends since working at home. Over half (55 percent) miss the camaraderie of a workplace.

Though working at home does appear to have some distractions and precipitate feelings of isolation for some of the sample, nevertheless, community life for some home-based workers seems to be strengthened. Over two-thirds of the respondents (69 percent) spend their leisure time close to home when they work at home. However, when these data are compared to reported amounts of time engaged in leisure activities and use of community services, no pattern emerges of active community use. Half of the respondents acknowledge that they know their neighbours better when they work at home, but only casually. Despite the mixed responses, it appears that working at home is perceived very positively in terms of reducing stress and maintaining household relationships and responsibilities while conducting work, and it is perceived somewhat positively in terms of enabling home-based workers to maintain social ties and activities outside the home.

Thirty-three (8 percent) of the respondents have disabilities. Most of these conditions, such as cancer, Crohn's disease, kidney failure, epilepsy, and mental health problems, do not require modifications to their home. Nevertheless, the respondents maintain that if they couldn't work at home they would not be able to continue to work, as it is difficult for them to function comfortably outside of their home. Some respondents who use wheelchairs or are confined to their beds have had to make modifications to their homes such as building ramps. They too acknowledge the importance of being able to work from their homes. Several respondents suffered from numerous respiratory infections while working in an office and have found the transition to their homes beneficial to their health. Other disabilities, such as back problems, migraines, and hearing impairment, have not been alleviated by working at home, but some people have addressed these problems by modifications to the equipment and furnishings in their home workspace.

This study revealed several distinct patterns. Telework has considerable positive impact on the job satisfaction of teleworking employees. Teleworkers, mainly because of the type of information-based work they do and their part-time status at home, rarely reported problems with integrating work into their home environment, and few reported problems with neighbours and regulatory bodies. Teleworkers in this sample generally believe that they have more control of their work when they work at home and, in turn, they are more positive about its execution. They especially appreciate the flexibility they have in their use of time and space over the course of the workday. They feel more efficient when they work at home and generally believe that they are highly productive. Most telework programs allow employees to work at home only part time. This also has an impact on teleworkers' satisfaction because they do not feel disassociated from the corporate culture.

Nevertheless, some teleworkers did report problems with "overwork" and inadequate work environments. Computer-related health problems, such as severe headaches and back pains, are often caused by ergonomically incorrect or badly positioned furniture. In recognition of these issues, minimum safety and ergonomic standards need to be maintained in the home for teleworkers.

While most workers want a more flexible work arrangement, working at home may not be the only work alternative. The home can be unsuitable as a workplace for many people because of spatial constraints and the lack of social contacts. Home-based employees may feel that opportunities for advancement are curtailed. Many employees are not self-motivators and cannot cope with managing their home and work responsibilities in the same environment. Though the survey results report a high degree of satisfaction with working at home among teleworkers, one-quarter of the sample, as will be shown in Chapter 7, is interested in an alternative work site close to home.

The self-employed segment of the home-based work population is more mixed in their job satisfaction. They work long hours, have few breaks, and often do several different jobs to make their livings. While they are often well educated and have extensive work experience, their incomes do not reflect this. They often report difficulty in mixing home and work activities.

The findings in this study generally reflect the experiences of skilled workers and professionals. Independent contractors and pieceworkers did not respond to this survey in appreciable numbers. Pieceworkers, many of whom are low-income and visible minority women, are reluctant to respond to surveys out of fear that government agencies may detect their work activities (and income), and often lack adequate language skills to respond. The findings demonstrate, however, that independent contractors work the most number of hours at home, are predominantly female, and generally have the lowest income. Their work is mainly routinized labour such as data processing. They are more likely to have to combine child care responsibilities with their work and tend to work in spaces that are used for other activities (see Chapter 6). Independent contractors are the least satisfied with their at-home work situation, due to the pressures of combining work and family responsibilities, inadequate work environments, and isolation.

This survey reflects the trend in Canada in the 1990s toward the polarization of good-job and bad-job sectors, which has created a widening earnings gap between the highly skilled and the unskilled. Statistics Canada (1999b) found that while the fastest-growing occupation group in percentage terms is the natural and applied sciences (including high technology), in absolute employment growth, sales and service jobs have increased the most (by nearly 500,000). Most of the increase in the science occupations has been in computer system analysts and programmers. Employment in these categories jumped 34 percent from 1989 to 1998, more than three times the overall employment growth. In sales and service the job creation has been in such positions as call centre workers.

On average in Canada, the high technology occupations earn the highest hourly wage, while the low-skilled workers earn the lowest hourly pay rate. There has also been a stark contrast in the occupations of men and women. Though there has been some movement by women to management occupations, and by men to sales and service, the "traditional" occupational structures of the genders have generally been maintained. The consequences of these trends will be further analyzed in the next chapter, which focuses on telework in Vancouver, British Columbia.

5
Localizing the Networked Economy: A Vancouver Case Study

Our societies are increasingly structured around a bipolar opposition between the Net and the Self (Castells 1996, 3).

In describing the networked economy, Castells (1996) analyzes how global economic interdependence is changing the relationship between the economy, the state, and society, the increasing flexibility and decentralization of corporate entities internally and networking between entities externally, and the individualization and instability of working relationships. Concurrently, there has been a rise in collective identities that challenge globalization. Information technologies have been instrumental in allowing the restructuring that has occurred. Computers, the software that runs on them, and the networks that connect them are the key enabling technologies of the new economy. IT is now a significant part of the US economy, about 11 percent of the GDP, but still less than 10 percent of the Canadian GDP.

While there is a tendency in the networked economy for the concentration of wealth and resources in "world class" cities, the decentring that is created by network linkages is precipitating "location-independent" relationships in peripheral cities such as Vancouver (Delaney 1994). Vancouver is distanced from its provincial and national context, most obviously culturally and increasingly economically, connected more to Hong Kong and Los Angeles than Ottawa. In comparison with most American cities it exhibits a fair degree of hybridity – the global and the local are intermingled. Rather than having a "New Spatial Apartheid" of segregation, in which the suburbs are predominantly white and middle-class and the central cities are populated by poor Hispanic, Asian, and African Americans, as in some American cities (M. Davis 1993), Vancouver and its suburbs are transcultural in their urbanism.

Vancouver's economy has uncoupled itself from its provincial hinterland and over half of the province's jobs are located in the city (C. Davis 1993). While Vancouver lacks major corporate headquarters, the employment base is dominated by the tertiary activities focused there, including tourism, retailing, management of Asian capital, high-technology industry, entertainment, advanced education, and health care. Given this diversified economy and Vancouver's attractiveness in terms of lifestyle (i.e., climate, education, recreation), Vancouver ranks fourth with Seattle in growth rate among North American metropolitan areas.

The rise in non-traditional work environments and flexible work arrangements corresponds to the changing economic landscape of the province.[1] While the largest proportion of firms (13 percent) located in the greater Vancouver area are in the forestry sector, the second-largest (10 percent) group is classified as high technology. There has been a 4.5 percent annual growth in high technology industry for the last several years, and the sector employs 46,000 people in British Columbia. While the IT industries are attracting the most attention, one of the fastest-growing industries in the export sector in the Vancouver area is the garment industry. There are 300 garment manufacturers in the province employing 7,000 workers. Of those, it is estimated that 1,500 are homeworkers working as sewers by the piece and often being paid below minimum wage. Generally, the companies headquartered in the greater Vancouver area are oriented to goods and services for the consumer marketplace (Greater Vancouver Regional District 1999).

Such factors as the low costs of labour and production compared to the United States and the high quality of life in Vancouver are attracting firms, though many relocate or launch an American corporate presence once they become established, because of the comparatively high Canadian taxation rate and the difficulty in finding qualified experienced personnel and capital for expansion. Of particular note, the film and television production industry has become a significant economic generator with substantial ancillary economic activity in pre- and postproduction activity. The new information technology economy that is being lauded in the popular press is also growing in Vancouver, and comprises a wide variety of goods and services including software developers, Web support companies, and e-commerce start-ups. The proliferation of IT companies means there are a number of teleworkers who work entirely on-line in informal work arrangements with their employers.

Telework in Vancouver

There has been a steady, significant increase in the number of home-based workers since the 1980s in Vancouver. In 1981 there were only 25,800 reported homeworkers in the Greater Vancouver Regional District (GVRD), but in 1991, 55,661 did paid work in the home. In British Columbia the 1996 census revealed that 8.2 percent of the total labour force, or 155,455 people, worked at home. Of those, 69,885 (45 percent) were located in the GVRD, a 25 percent increase from 1991. The number of part-time workers has also increased. In 1989, they were 13 percent of the labour force; in 1999, they constituted 18 percent.

Self-employed entrepreneurs still constitute the largest percentage of teleworkers. Formal telework programs, as described in the previous chapter, do not seem as prevalent. Though the Web sites of various government and non-governmental organizations and telework consultants – e.g., Canadian Telework Association, Telework Canada, Telework International – listed a number of organizations in the greater Vancouver area with telework programs, research reveals that none are currently implemented. These Web sites strongly advocate this changing workplace but often provide inaccurate data about the realities of telework.

Out of the eight major companies contacted in the GVRD, none currently have formal telework programs though three allow flexibility in where employees can work. At a leading software engineering firm based in Richmond, employees work individually or in groups on projects and are free to do work where they see fit given the context of the project. A representative of the company described the arrangement as just "whenever, wherever." At another high technology firm based in Burnaby, if people are working from home it is usually on an informal basis privately arranged with their supervisors and not formally endorsed by the company.

Several programs were discontinued or never implemented. The dominant telecommunications carrier in Vancouver initiated a telework pilot program in the early 1990s but subsequently terminated it due to corporate restructuring and downsizing, though a satellite office in an outlying suburb was still in operation in 1995. An energy-producing company had proposed a program several years earlier that was abandoned due to financial restructuring and cutbacks in the company. Another provincial energy-producing corporation had plans to introduce telework stopped by their unions. The union governing the "internal workers" (i.e., the support staff) was the most vocal critic.

The trend toward independent contractors doing data entry also appears to be lessening, because these workers' homes lack sophisticated computer technology and confidentiality of records cannot be maintained in unsupervised surroundings. Research on Vancouver-area firms employing data entry workers revealed that only a few actually employ home-based workers. Canada Employment, temporary placement, and secretarial support agencies do not handle home-based workers, and representatives of the private agencies even had trouble understanding what the term meant. It would appear that telework and home-based employment have not been integrated into mainstream work options if the most likely places to seek this work have no connection to it whatsoever.

An employment counsellor at the Human Resources Development Canada (HRDC) office in Vancouver indicated that home-based work is not as common it is perceived to be in the popular press. Often it is difficult for low-skilled workers to keep pace with the rapid changes occurring in computer technology, and they cannot afford to constantly upgrade as new innovations are introduced. There is also a fear, whether perceived or real, that data can be intercepted or read en route from the home-based worker's computer to the receiving office. Interviews conducted with data entry workers supported these conclusions. They had quite obsolescent computers (e.g., 386 PCs) and were still sending disks back and forth to their employers rather than sending data by e-mail, since they did not have that capability on their machines and their employers preferred this arrangement.

Another major factor that has contributed to the reduction in data entry work is the introduction of scanners in the mid-1990s. Now that data can be entered directly into the computer without rekeying, data entry has become a "dying field," in the words of one home-based data entry operator. The rapid ascent and descent of this work illustrates "dis-intermediation," the elimination of the middle person that has occurred as technologies increasingly allow networking between machines. Data entry work was quite significant in the 1980s, though critics were concerned about the increased automation of tasks and decline in the quality of work, especially for women. These concerns have become inconsequential in the early twenty-first century as these jobs become obsolete. It appears that data processing is not steady work, so many women who do it have other sources of income and take on data processing for additional income when they can get it. In addition, some workers who are employed in this area work one or two days at home in an informal arrangement.

The busy owner of a medical transcription service in the Vancouver area, who is on call eighteen hours per day, seven days per week, describes why she does not hire home-based workers. Her physician clients are concerned about breaches of confidentiality, such as patient files on kitchen tables when people drop by. The nature of her company's work often requires a fast turnaround, and women working at home are often fitting work into family and household tasks. The work can't be reliably, consistently completed on time at home. In addition, her clients can have unpredictable schedules. For example, if surgery ends at 10:30 p.m. and the file must be processed and ready for the next physician the next morning, often a home-based worker cannot accommodate this "just in time" demand. The service owner cannot afford the time to phone around until she finds someone willing to take the job on that night. As she concludes, "Flexible, high quality, and quick service is the edge of competition in my field, and since there's a limited market the competition is tough."

Home-based Work Trends
The 1996 census revealed a number of interesting patterns about where homeworkers reside (Figure 5.1). West Vancouver, one of the wealthiest municipalities in Canada, has the largest percentage of at-home workers in

Figure 5.1

Percentage of people working at home, Greater Vancouver Regional District, 1996

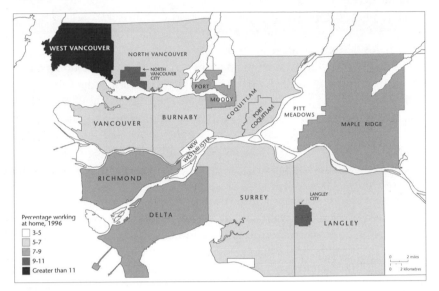

Note: The number of people in the GVRD working at home in 1996 totalled 69,885, out of 844,400 people employed. In 1991, 55,661 out of 822,080 worked at home; in 1981, 25,800 out of 601,515 worked at home.
Source: Data from Statistics Canada (1997).

the GVRD (17.6 percent of the total municipal labour force) while the lowest percentage (4.4 percent) of home-based workers is in a predominantly low-income community, New Westminster. The prevalence of home-based work in West Vancouver can be attributed to "knowledge workers." There appears to be a high incidence of professionals in the community who are opting to work out of their homes: people in the design professions, lawyers, and those in the financial sector, such as accountants and stockbrokers.

Surrey, a historically low-income community that is becoming more mixed income, had a dramatic decrease in home-based employment activity from 1991 to 1996, with a 5 percent total reduction and a 20 percent decrease in male participation (Table 5.1). In most other communities there seems to be a steady increase in this activity, especially for women. The other exceptions are Langley Township (a semi-rural area that is undergoing significant suburban growth) and North Vancouver District. The largest percentage of growth between 1991 and 1996 in female home-based work occurred in Port Moody (34.3 percent growth) followed by Burnaby (23.4 percent growth), both predominantly middle-class communities. West Vancouver also had a significant increase in female home-based work activity (19.6

percent), but the real growth in this community was among men (26.2 percent). It appears from a comparison of 1991 and 1996 census data that home-based work activity is growing more in the high- and middle-income communities than in the low-income ones.

While more comprehensive data collection by Statistics Canada has resulted in a better understanding of non-traditional work settings, the extent of this activity is still ambiguous. As with the difficulty in obtaining national reported statistics, municipalities do not have a clear understanding of the amount of home-based work activity in their jurisdictions. Many home-based businesses are not visible and are operating without the knowledge of the municipality. As most current zoning prohibits home occupations, home-based business operators would rather not be licensed. In addition, teleworkers working as employees are not included in reported data on home-based work activity, as employed workers are not considered to be operating a business from their homes.

The most common types of home businesses operating in the municipalities of the GVRD are professional and secretarial services, general building and trucking contractors, accountants, and consultants. The second most common forms of home businesses are daycares, mobile services, manufacturers' distributor agents, wholesale dealers, and import/exporters, followed by clothing manufacturers, hairdressers, janitor services, small group education and training, and arts and crafts. The types of home occupations are changing as information and communication technology makes more home occupations viable in all aspects of the job market. Consequently, more professionals are now working at home. In addition, there are more home crafts licences. However, most municipalities do not clearly define what constitutes a home occupation.

With the decentralization of employment to suburbs in the greater Vancouver area, policy makers are starting to view home-based business as important to the financial growth of outlying communities, and a way to reduce provincial costs associated with transportation. Home occupations, if carefully regulated to balance entrepreneurs' desires and community goals, do not have to be a contentious issue. Problems arise when bylaws and regulations are too general or too restrictive. One municipality resolved these problems by taking direction from a local community group that formed to review the bylaws pertaining to home-based businesses. Though the process was lengthy, it resulted in changes to policies and regulations.

Telework in Daily Life
The information sector, as the analysis of employment patterns has demonstrated, is one of the significant engines driving the Vancouver economy. How this is played out "on the ground" – in the daily life patterns of people – reflects an off-balance work culture where in many cases work has all but

Table 5.1

Percentage of municipal labour force using their homes as usual place of work in municipalities of the Greater Vancouver Regional District

Municipality	1981			1991			1996			% growth 1991-6		
	M	F	T	M	F	T	M	F	T	M	F	T
Burnaby	1.9	3.3	2.6	4.6	4.7	4.7	4.7	5.8	5.3	2.2	23.4	12.8
Coquitlam	2.2	3.5	2.8	5.3	6.0	5.7	5.4	6.9	6.2	1.9	15.0	8.8
Delta	3.5	5.1	4.3	5.2	7.1	6.2	5.8	8.6	7.2	11.5	21.1	16.3
Langley City	–	–	–	9.9	12.3	11.1	9.1	12.9	11.0	-8.0	4.9	-1.0
Langley Township	–	–	–	4.0	7.5	5.8	4.7	7.1	5.9	17.5	-6.7	1.7
Maple Ridge	–	–	–	–	–	5.5	9.3	7.4		–	–	–
New Westminister	2.5	2.9	2.7	4.1	4.6	4.4	3.9	4.9	4.4	-5.0	6.5	0.0
North Vancouver City	3.8	5.5	4.7	8.4	9.2	8.8	8.2	10.9	9.6	-3.0	18.5	9.1
North Vancouver District	3.0	3.0	3.0	6.1	5.5	5.8	6.4	5.4	5.9	4.9	-1.0	10.2
Pitt Meadows	–	–	–	–	–	–	7.1	1.2	4.2	–	–	–
Port Coquitlam	1.7	2.0	1.9	4.6	6.1	5.4	5.1	7.1	6.1	10.9	16.4	13.0
Port Moody	2.9	3.9	3.4	5.8	6.7	6.3	6.5	9.0	7.8	12.1	34.3	23.8
Richmond	3.1	3.5	3.3	5.7	6.8	6.3	6.6	7.8	7.2	15.8	14.7	14.3
Surrey	–	–	–	6.3	7.6	7.0	5.1	8.3	6.7	-20.0	9.2	-5.0
Vancouver	–	–	–	6.0	5.8	5.9	6.3	6.8	6.6	5.0	17.2	11.9
West Vancouver	–	–	–	12.6	16.2	14.3	15.9	19.4	17.6	26.2	19.6	23.1

Note: 1991 place of work data are not directly comparable to 1996 data, as the options in 1996 were changed to include "no usual place of work." This may have resulted in an underestimation of the growth of home-based employment between 1991 and 1996.
Source: Statistics Canada (1981, 1991, 1997).

completely eclipsed personal life, and productivity at work takes precedence over time spent with family. Instead of the nine to five routine, work is conducted almost anywhere, from early in the morning to late at night (five to nine), and scheduling of "down time" is predicated on the work schedule. Often work comprises not just one paid employment, but several jobs done concurrently in order to support a family. Given shrinking benefits, downsized corporations, and insecure employment, it is the imperative of sheer economic survival that propels most workers. Telework and other forms of home-based employment fit into these survival strategies.

As Denise, a resident of Vancouver, aptly articulates, "I have given up trying to reach a balance. I recognize that all I can do now is juggle." Denise has three jobs: she is a therapist and editor of a newsletter on counselling psychology, and with a partner she is developing a Web site offering information, referrals, and counselling services. She recognizes that she has another full-time job – taking care and managing the schedule of her five-year-old daughter. Denise sees clients in the evening in a separate suite in her house, and afterward she often works on her projects at her computer until 2:00 a.m.

While Denise is stretched in terms of time and in balancing her various demands, she is also a trained professional with a number of options open to her. Others do not have that flexibility. Monica, forty-nine, acknowledges that she is in a precarious situation. She and her family are recent immigrants from Hong Kong. After giving up a comfortable life in Asia to pursue opportunities here, she and her husband are finding it almost impossible to maintain a decent standard of living in Richmond, the suburban community adjacent to Vancouver where they reside. Her husband has two jobs, working fifteen or sixteen hours a day. He rarely gets to see his sixteen-year-old son. Very worried about their son's performance in school and the lack of parental supervision because of her husband's heavy workload, but also realizing that they needed extra income, Monica started to do data entry work at home. She liked the arrangement because "It allow[ed] me to work around the needs of my son." She would work during the day when he was at school, and after dinner worked again until 11:00 or 11:30. As she said, "Sometimes I feel I work too much, feel tired, and cannot keep the house perfect." She used to work on average six or seven hours a day, four days per week, but that ended two months before she was interviewed because of lack of work. She is now working from 7:00 p.m. to 3:00 a.m. for the same company, but in their corporate headquarters. She is highly concerned about the consequences of this work arrangement for her son.

Amanda, forty-one, also acknowledges her limited options. She has been doing data entry for the last twenty years and has done it at home for the last six. She has a day job registering liens in a bank in addition to her home-based work. Working forty hours per week at her regular job, she puts in another twenty hours (four or five hours per day) most weeks at home. She works on the weekends as well, trying to put in eight hours a day. Sometimes she works through the night. On a typical day she starts work at her day job at 6:00 a.m. and gets home at 2:30. She rests and then does data entry from 4:00 until about 10:00. When asked about when she does household chores, she acknowledges that it is "whenever I get spare time." She rarely has time for leisure activities. According to her, "The extra time I do have, I rest." She knows that she works too much, but according to her, "That is how life is these days for most people." She has to work to pay debts. She makes less than $29,000 per year.

Amanda lives alone. She used to rent a three-bedroom townhouse in Richmond and converted one of the bedrooms to an office. It was sold and she subsequently moved to a one-bedroom apartment in Surrey, where she works in her living room. Her computer is a PC 386, now considered antiquated. She does not have a modem and instead picks up her work and returns disks to her employer in ASCII format. She uses two programs for her work, Access and Promark. She is paid 9 to 12 cents per line, based on the difficulty of the document. When she works at home she "just works, I don't really get to know my neighbours that well."

Her situation changed a few months before the interview because there was no data entry work. She was laid off her data entry job and she is now setting up jobs at night on high speed laser line printers in a company's office, working about six hours a day on top of her regular employment. She would prefer data entry work and would ideally like to have her own business and employ other people, but admits that "there is not much data entry – it comes and goes."

Amanda prefers to be her "own boss." She does not like office cultures, where she finds that people are not motivated and do not work hard. She has no problem working hard, though she recognizes the toll it is taking on her body in neck strain and back pain. She is very focused when she is working at home and prides herself on her self-discipline: "I am there by myself I have to stay focused, motivate myself to sit there and do the job. I am kind of a control person. When working by the piece I have to be fast and accurate ... takes a lot of discipline to keep speed up ... if take breaks often speed goes up." Amanda has some postsecondary education in accounting but it is incomplete. She would like to take some computer courses but again concedes that time is a problem, since so many of her waking moments are spent working.

Katrina, thirty-four, has been doing data entry work for just over twelve years. Originally she worked at the office of a data entry company in Victoria, but in August 1998 her husband was transferred 100 kilometres north to Nanaimo. Because the company did not want to lose her, they agreed to let her continue working in her home. She had to buy her own computer.

She gets no benefits and is paid on an hourly rate. Twice a month she submits a time sheet. Her work output is monitored by a program, Data Entry Emulator, that measures her keystrokes. The computer programming work can not be measured in this manner and her time sheet entries for this work are based "purely on trust." The company phones every day at 2:00 to see if she wants work for the next day. If she does, the documents are delivered to her by courier and she usually returns her completed work on disk the same way. Some of her completed work is sent by e-mail, but only to specific clients. She has more responsibility since working at home: in addition to data entry, she now also does computer programming and troubleshooting. Once every three months she goes to her office in Victoria, and her employer comes to her home once a month.

She works on average forty hours per week while also trying to manage child care. She prefers to be at home because of her children: a four-year-old son at home, a seven-year-old daughter in school, and a baby on the way. However, without the child care arrangements she does have she wouldn't be able to do her work. Though her son goes to preschool, he is at home with her the rest of the day. This arrangement "works out fine. I work around his schedule – early in the morning, at night, and on weekends when my husband is at home." She is up by 5:00 a.m. and works when everyone is sleeping. She gets her daughter up by 8:00 and gets her ready for school. She is in a carpool, so only has to drive her daughter to school once a week. After her daughter has left she continues working. Her son plays or watches a movie while she works in a nook off their kitchen in their three-bedroom home. At 11:30 she takes her son to his preschool, then works until 2:30 when both her children return home. She has a short break and then works until 5:00. At 6:00 her husband comes home and they have dinner. If her husband is at home during the evening she works until she goes to bed at 10:00. She estimates that she works about 50 percent of the time at night and during the weekend. Their total household income is around $65,000, half of which is from her home-based work.

Wendy, thirty-seven, has also been doing data entry work for just over twelve years, but during that time completed her undergraduate degree. She is currently a full-time medical student, doing data entry six hours per week, but while she was doing her undergraduate degree in the late 1980s she operated a large data entry business in Richmond, BC, employing twenty women out of their homes. She started working at home because it was an ideal way to combine family, school, and work. She recognizes the limited opportunities in this field, mainly due to the advent of scanners in the mid-1990s that are replacing human data input. However, she has a number of well-established clients and still makes a good income. Because she is very fast, she can make as much as $250 in two hours of work, working by the piece. She uses a new Acer computer, but while she has the technological capabilities to send her completed work by e-mail, her clients need a hard copy of her work and still want a disk. Her total household income is over $85,000 a year.

Wendy has a daughter, twelve, and a husband, forty-seven, and they live in a large five-bedroom home in Richmond. She has converted one of the bedrooms into her office. She cannot describe her daily schedule as she has "no rudimentary type of lifestyle. It is not like your average nine to five. It is like describing a housewife's schedule. When do you break for lunch? When you have a boss some people need to be told. When you are a home-based worker there is no system. Why do you have to have lunch at noon or start at nine? I have no set schedule. It is almost like there is no clock." She usually wakes up at 4:00 a.m. and either studies or works. Her daughter gets up on her own and makes her own breakfast. She works through the day and rarely breaks for lunch. She stops work whenever she has to do something else. She often works until 11:00 or midnight but sometimes falls asleep at 6:00 p.m. As she stresses, "I am not a wonder woman."

Like Amanda, Wendy emphasizes the need for self-motivation in this type of work: "You definitely have to be a self-motivator. There are so many distractions. It takes a different sort of person. It is not for everybody because of personality. If you are the kind of person that needs to be told everything it won't work for you. I know some lawyers who were self-employed and went to the golf course once too many times. They went bankrupt and are now working for someone else." She is, however, very enthusiastic about this work arrangement: "Home-based work is the only way to go. You don't have anyone at the helm having control over you. I am teaching my daughter a different attitude towards work. I say to her you must have it done today. I don't care when it is done but you must have it done."

Even those who at first appearance seem to be the "haves" in the new economy struggle constantly with maintaining a balance in their lives. Tim, thirty-one, is living the flexible work life but constantly trying to juggle the demands of his family and his job as the creative director of a small computer graphics company in Vancouver that produces graphics for film, television, and new media products such as Web sites and CD-ROMs. As a "knowledge worker," his work is "constant challenges, a lot of variety, very fast turnaround time, intense environment up against deadlines, very creative. I can do what I want – come and go as I please."

Tim, who has some postsecondary education in art, initially worked in graphics support for a large corporation. Feeling the need for a challenge, he taught himself computer graphics and left his job. He worked out of his home as a freelance consultant for a variety of companies, including his present employer. As he terms it, "We were dating before getting married." They couldn't afford to hire him full time and the company is still very volatile. It has no benefits package. He worked as a consultant for the first ten months of his association with his company. For the first three months he worked entirely at home, after that, two days at home, three days in the office. He now works mainly in the office but usually brings work home in the evenings and often stays home to work when there are particularly pressing deadlines. He started on the payroll in July 1999, and his goal is to own part of the company. His total household income is over $85,000 per year with his work contributing two-thirds of that income.

When he was working at home full time, he worked on average fifty-two hours per week, not unlike his present work situation. However, he finds that at home his workday spreads out through the day. He has two young daughters, four years old and ten months old, who demand his time, and he tries not to work when they are at home and awake. He is actively involved in household chores and child care. He lives in a two-bedroom apartment in faculty/staff housing on a university campus, where half of the living room has been converted into his office. His wife was working as a laboratory technician at the university but is now at home with the baby. When Tim worked at home full time he would wait until after taking his older daughter to daycare around 9:00 a.m. to begin work. He would call his office to check in and then work until around noon. He would go for a walk, and often go out for lunch, to get out of his home. He would then work until he had to pick up his daughter at around 5:00 p.m. He would definitely work after her bedtime, usually until 1:30 or 2:00 a.m.

Now that he is working from an office he finds that his time is more structured, but because of the extra demands of a new baby, he is still stretched for time. He is at work by 9:00 or 9:30 a.m. and leaves for home by 6:00 p.m. He then spends time playing with his children, getting them ready for bed, doing household chores, and relaxing. He often works on his home computer from 9:30 until past midnight.

Tim is conflicted about the benefits of telework. On one hand, "It makes me feel more independent from my job. Feels a lot like contracting. I am my own identity, a 'hired gun feeling.' Makes me feel like I am a free agent doing it myself. I feel I have more control over my work life." On the other hand, it is very isolating: "I was starved for adult companionship. It is very lonely. I sought out full-time work for companionship. I have good friends at work. The bulk of my social interactions take place at work. I don't like the pressure of working on my own." It is also difficult to build up a creative team remotely: "Working from home removes me from the team. As an administrator I prefer to have my creative staff working close to me. It is a matter of control, but it also fosters a team environment." His preference is to have a much more structured day: "I prefer to work nine to five and then stop. I don't like that I have to work in the evenings. It detracts from what

I should be doing with my wife and kids." He acknowledges that he can work equally well at home or in an office because it is all electronically based: "It makes little difference where I work. I need a computer server that I can connect to and I transfer all of my work by FTP (file-transfer protocol) to my work associates and clients. All of my work is done digitally. There is no paper."

Tim's work is almost all done for American companies based in Los Angeles. Though his company is very competitive and produces good work, it cannot get around the dominance of the American market, where it is a "bit player."

Anna, thirty-nine, is also a "knowledge worker" selling her services internationally. Born in China, she has a PhD in physics from an American university and since 1995 has consulted on Chinese Web-based initiatives for various American scientific publications. She is based in Vancouver because her husband is a postdoctoral fellow at a university there. They have a five-year-old daughter. Her work entails setting up server sites to support the dissemination of information and updating the servers. Her clients are in Boston and Washington, DC, so she rarely sees them. They communicate almost solely by e-mail and several times a year she goes to China to work on her projects there. Her computer skills are self-taught. She is working more than half time at her work and rarely has time for housework.

Her work schedule is much like that of a graduate student. While she works during the day, her most productive time is at night after her daughter has gone to bed, from 9:00 p.m. until after midnight. As she acknowledges, "I always had flexibility in my work. I never had a structured life. It is isolating though." She is not a social person. Since graduate school she has not made many friends. She also feels out of place. As she emphasizes, "I always feel marginal. I am not quite Chinese because I studied in the US but not quite North American because I am Chinese." She typifies herself as very self-motivated, otherwise it would be very difficult to do this type of work remotely.

Derek, twenty-nine, a technology support consultant, has worked in the paid workforce for only seven years but recognizes its inherent tensions. His work entails suggesting solutions for problems that people have regarding the Internet, networking, and PC platforms. He views himself more as a generalist than a specialist. Half of his work deals with implementation and the rest concerns solutions, analysis, and design. He has worked for multimedia, shipping, manufacturing, and high technology companies with workforces ranging from three to 100 people on ongoing maintenance and software troubleshooting. One of his current contracts is in Los Angeles, helping a shipping company to streamline its operations and expand its information systems to take advantage of the Internet. He is the designated point person for a team of people from Vancouver, Seattle, and Los Angeles.

He bills by the hour or sometimes on a turnkey basis – a lump sum. His and his wife's combined income is over $85,000 a year.

Though born and educated in Vancouver, after receiving his computer engineering degree he went to work for a large software company in Seattle for two years. He found that "working in Seattle was very stressful. I wanted a bit more balance in my life. I am not lazy but no one wants to work all of the time. I found myself spending all of my time in front of the screen. My eyeglass prescription was getting stronger and stronger. I was only getting home [to Vancouver] once or twice a month." In order to change his situation he quit his job, moved back to Vancouver, got married, and started to do contracts from his home, initially only part time while he worked at another full-time job. He has been fully self-employed since July 1999. Derek comes from a Chinese-Canadian family. While his wife understands his work choice, and his brother and sister are also self-employed consultants, his mother is mystified by what her son does beyond working with computers.

He is very pleased with his new work arrangement. The "lifestyle choice" is a big factor in his satisfaction. As Derek states, "I have more freedom. I am my own boss. I am away from an office environment. I am in control of my own time and in a place where I have more options and doing what I want. I am not into empire-building. I can sustain a certain level [of work] that I am comfortable with." He works about twenty-five hours a week in his home and another five in other work locations, spends seven hours per week on household chores, and still has five or six hours for leisure activities. He enjoys golf and hiking, which he never had time for when he worked in a downtown office and returned home at 7:00 or 8:00 at night. He now gets up late, about 9:00 a.m., and immediately checks his e-mail and responds to urgent business. He then showers and has breakfast, then works or does chores or errands. He starts work by checking the electronic personal organizer that tracks his various work commitments. At 1:30 he has an elaborate lunch lasting one or two hours and then either works or goes out golfing or some other activity. He has supper with his wife and then starts working again around 9:30 p.m. until midnight. Beyond that, he will usually work another couple of hours but describes that time as "more leisurely, more surfing the net than working, though still being productive." When in front of the computer he tries to institute regular breaks. He shuts down his computer at 2:00 a.m. and goes to bed.

Derek has learned how to set clear boundaries between his work and domestic life. He lives in a four-bedroom home, and has converted the bedroom that is most physically separate from the domestic functions and closest to the main entrance into his office. He has a sign with three round faces and a movable arrow on his door – one with a smiling upturned mouth indicating "fine, interrupt," a serious face with a downturned mouth indicating "serious work mode, don't interrupt," and a neutral face with a straight mouth indicating "maybe interrupt." He minimizes personal calls during the day, but he does admit that he has "time management issues." Sometimes working hours are lost in frivolous pursuits.

Derek's work is highly dependent on e-mail and the Internet. He spends about 20 percent of his time sending and receiving e-mails and 30 to 40 percent in Internet programming, document creation, and using database tools on his word processor. While he always has initial face-to-face meetings with clients, he finds that e-mail is the best form of communication as work continues, because it leaves an audit trail. Agreements are made in writing, rather than just verbally, and decisions can be made in a few words. In addition, "E-mail disciplines you. Your client is waiting for a response and the ball is in your court. It is more accessible and more discrete. It is easier to connect with people during the day via e-mail than the telephone. E-mail responses from clients and coworkers give me insight into how much they know, how busy they are. I can cultivate a deeper relationship with them. But e-mail communication can't convey emotion very well. Things sometimes get completely blown out of proportion if messages are perceived differently than their intended purpose."

At first glance the life of Karen, forty-six, would almost seem idyllic and worth the envy of the teleworkers described above. Living on acreage on an island an easy commute from Vancouver in a house she shares with her husband and three sons aged eleven, nine, and six, she concedes that she works at "too many things." She manages a large commercial Web site that involves constant updating, provides administrative support and research to her husband's at-home software and telecommunications consultancy, writes fiction, and imports rugs from North Africa. For her Web-based work she is on contract – a stipend based on the percentage of revenues. She has an MA in communications and has worked for ten years in these occupations. She started to work at home because she wanted to have time to write fiction.

She works over thirty hours per week in her various occupations, spends at least fifteen hours doing housework, and rarely has time for leisure activities or volunteer work. As she typifies her workday, "I squeeze things in here and there. I want to be with my family most of the time. I can't do concentrated in-depth work over an extended period of time. I can do it in little bits of time and when it arises I jump in and get a few things done. I have a 'hunting and gathering' strategy of work. That is why I got involved in the Web site. I thought I would have snatches of time to work on it." She gets up around 7:00 a.m., always checks her e-mail and then gets her children ready for school. She maybe works about an hour in the morning. In the afternoon she works two hours more. In the evening, after her children go to bed, she always works from 10:00 p.m. to about 1:00 a.m. She has to put in more hours when she has to update the Web site, a frequent occurrence.

She knows that her children resent her work; they perceive that it prevents her from fully participating in their activities. She also knows that if she were still living in Vancouver she would probably be working twelve hours a day.

Karen recognizes that telecommunications and information technologies have allowed her the flexibility to work on her island. The house has three telephone lines, one for her husband's work, one for their fax and modem, and one for their residential line. Karen has noticed that the use of e-mail to conduct most of her work has changed expectations. "There is an increasing expectation of immediate reaction. If I get an e-mail and don't send an immediate response, clients and associates wonder what is happening. With e-mail you have to be constantly 'there.' It makes you accessible all of the time. I spend a lot of time in that environment [e-mail]. But that cyber-connection seldom leads to friends."

She finds that telephone communication regarding work can be worse: "No boundaries ... if pick up call lose any boundaries [between home and work]." At least with e-mail she can respond when she wants to. She prefers e-mail meetings because she "can control timing and keep things on topic." Karen recognizes, "I can't relax at home because that is where I work. I always have to do chores. I would like to have a separate workspace. I would like to separate my home and work environment. I need solitude in my work and I can't get that in my home."

On the island where Karen lives there are many other home-based workers: professionals, artists, knowledge workers. It is "a social, real, functioning community." There is a shopping centre close to where she lives that has a business centre with office support such as photocopying and faxing, as well as offices for rent. This is where Karen would like to move her office. Ideally she would like to maintain duplicate workspaces at home and in an office.

The Cyborg Worker

These descriptions of teleworkers illustrate several aspects of the new economy. The traditional measure of work, the workday, is disappearing. The idea of the eight-hour day – a creation of the industrial revolution – doesn't fit in a world where work is measured by tasks rather than by the clock. For some people this means more freedom to self-define what a work schedule constitutes. For others, it means fitting work into other activities, as Karen does, squeezing work in "here and there." For still others it means the licence to "overwork." The workday, nevertheless, is an anachronistic way to define output and rarely reflects the experiences of workers. The speed with which work is conducted is the main reason the workday is redundant. The new economy is based on fast turnaround and the work experience described above certainly reflects that. Data entry workers have daily deadlines to meet. Knowledge workers have to respond immediately to requests. To fulfil these work commitments often means working over-fifteen-hour days spread out through the day, evening, and night, whenever there is time to work.

Another aspect of the new economy is the emphasis on communication. All four of the knowledge workers profiled are directly involved in creating, managing, and disseminating information. The skill that is sought after in knowledge workers is not just technical, but their skill in making decisions. As decision makers managing their own projects, they are rarely satisfied with a traditional employer-employee relationship. The emphasis that all of those profiled placed in being their "own bosses" reflects not only a personality trait, but the nature of work and work organization in an economy that rewards independent thinkers.

However, as these profiles demonstrate, this profound redefinition of work life has not occurred just because technologies enabled it, but because there has been a major shift in household priorities. Women are not just working to be fulfilled; they are working to maintain the standard of living of their families. While men have to put an enormous amount of their energy into work because of increased expectations and economic imperatives, they are still trying to maintain their family connection. The toll this redefinition of work is taking on families is a direct consequence of the lack of societal support for these changing identities.

As demonstrated by the descriptions of data entry pieceworkers losing their jobs as technological innovations have eliminated their work, de-skilling and job loss is occurring for low-skilled, low-income workers. Their work is becoming redundant and their job prospects are bleak. Those who remain employed have very little control over the terms of their work and their use of technologies, and often have their performance monitored via computer. The difference between the knowledge workers and the pieceworkers can be typified by their relationship to computer technologies: the knowledge workers work *with* computers, but the pieceworkers work *for* computers.

The knowledge workers, though they have a considerable amount of flexibility in where and when work can be conducted, are highly dependent on the global economy. All of the workers profiled work on projects that originate in other countries, mainly the United States. Although technological capabilities make geographic distance redundant, however, it does not surmount cultural differences. As Tim emphasized, his company is not seen as a "major player" in the United States because it is Canadian. In other words, technological capabilities and skills can not overcome cultural biases and economic imperatives. While decentralization of tasks and work location within corporations may be the inevitable consequence of an information economy, economic power in the networked economy remains concentrated.

The strong emphasis that all of the respondents placed on self-discipline reflects a strategy to maintain dignity and control in a work world that is increasingly becoming "out of control." These workers take pride in their work and in their ability to function as autonomous beings. Their dexterity in manoeuvring through the complexity of their work and home lives, even in the most arduous conditions, demonstrates a capacity for adaptation and innovation in the new workforce. In their habits of daily life they are resisting the totalizing forces of the new work reality.

Loneliness and isolation are the consequences of this excessive importance placed on the "self." There is no support network to buffer the hard times. In communities where there are strong community ties, individuals are sustained in difficult circumstances. The government took on some of this role through its "social safety net" of universal health care and other essential services, and corporations continued this role through their benefits packages. Progressive corporations recognize the importance of team-building to increase productivity and, as a by-product, foster "community." Home-based workers have few of these safety nets and do not even seem to expect them. Except for Tim, who recognizes the importance of team-building, the rest viewed the camaraderie in office environments negatively. "Community" and what it can offer is not a significant presence in their lives, with the exception of Karen, who lives in a strongly identified community.

What is the new identity that is being created by the networked economy? Not only is it a blurring of work and domestic activities, as described in other chapters, it is also a blurring of the distinctions between human and technological functions. The cyborg (half-human, half-machine) image comes to mind. Technologies can extend human capabilities, but conversely humans can be susceptible to being extensions of machines. This is certainly true of data entry work, which entails feeding (inputting) data into computers to be processed. This work has become superfluous because new technologies have developed that allow computers to "talk" to each other.

The cyborg image also reflects the increasing use of technologies not just as tools but as "immersive" environments. We not only immerse ourselves in the computing environment, we become inextricably dependent on and connected to its functioning. If knowledge workers did not have continuous access to the Internet and e-mail, their work output would effectively stop. As a consequence of this interweaving between the technological and the human, we are increasingly being watched by the new technologies. Data entry and call centre workers can be effectively and continuously monitored when they work. Knowledge workers, while not subject to such close scrutiny, are still effectively observed on the Internet in their patterns of use and preferences.

This chapter has analyzed the consequences of the networked economy in a particular locale, Vancouver, BC. The data reveal a bifurcated workforce made up of highly skilled, highly paid knowledge workers who are part of this economy and deriving benefits from it, and low-skilled, low-paid piece-workers who are being made redundant by new technologies. The "dis-intermediation" that is affecting this workforce is resulting in job loss and precarious employment opportunities. That women are particularly vulnerable is not surprising, given the multiple demands they have on their time and resources. The next two chapters will discuss how the use of space in the home and community further reinforces gender roles.

6

"I Don't Have a Home, I Live in My Office": Transformations in the Spaces of Daily Life

Speculation abounds about telework's possible impact on community stability and energy expenditure. Writers in the popular press hypothesize that working at home will deepen face-to-face and emotional relationships in both the home and the neighbourhood, mitigating the increasingly impersonal relationships outside these environments. Telework will increase community stability because people will not have to move if they change jobs; they will simply have to network with a different computer. Such a home-centred society will reduce the expenditure of energy for commuting and lead to energy decentralization, because energy will no longer be required for high-rise office buildings. As people increasingly work from home and other nontraditional workplaces, traditional residential communities that exclude work opportunities will no longer meet the full range of needs of their residents, and new and different urban forms and services will be required.

While these speculations have major implications for the organization of the home and community spheres, there is no conclusive empirical evidence that the use of residential neighbourhoods intensifies when people work at home. The findings from this research are mixed at best and do not support the contention that telework will reinvigorate home and neighbourhood life and activity. However, it should be recognized that these findings are within the context of homes and neighbourhoods as they exist. Active community development interventions and new forms of residential neighbourhood design that facilitate more locally based activities, such as walking to shops and recreation, may lead to different outcomes.

How teleworkers use their spaces for work needs to be understood in the context of the division of labour in the home and the meaning attached to social spaces. Homes and communities are the physical expressions of sociocultural beliefs and practices (Rapoport 1969). Spatial arrangements of homes reflect and reinforce the existing gender, race, and class relations in any given society and contribute to power differentials (Weisman 1992).

Socially constructed relations of power and hierarchy are reflected in people's organization of home spaces. For example, spatial patterns in a household with gender-egalitarian behaviour would differ from those in one with more rigidly defined gender roles. The analysis in the preceding chapters reveals that the introduction of telework in the home does not necessarily change the division of labour. Women are still the predominant caregivers and household maintainers and managers, regardless of where they do their paid work. The evidence on decisions regarding the use of space in the home bears out this conclusion.

From the profiles chronicled in the preceding chapters, several tendencies emerge, which correspond to the priorities of teleworkers. Because they have trouble defining themselves as workers, home-based workers whose primary priority was the care of their families (almost all women) usually work in spaces that are used for other activities. In contrast, those who make work their primary focus tend to define their workspaces more clearly. Often, however, work dominates the home environment. Gender, household status, and financial resources affect these tendencies.

Jenny, thirty-two, a medical transcriptionist, is one of the teleworkers with domestic priorities. The mother of two school-age children, she lives in a comfortable suburban house in Hercules, a suburb of San Francisco. She has worked at home for four years because she wanted to be with her family more. She works on average every weekday, twenty-five hours a week. Her office is a corner of the tiny bedroom that she shares with her husband. Jenny chose to work in her bedroom because she didn't want to intrude on any of the public areas of the home. She felt that if she worked in the family room, her children wouldn't accept that she was working. Her husband, an engineer, built her a desk, but he has brought work home and his manuals have invaded her desk space. Now, she is left with a small corner of the desk for her computer. Because she doesn't have a telephone outlet in her bedroom, a long wire connects her modem with the kitchen telephone. She acknowledges, "My children resent that I work, and my husband likes the extra money I am making but expects me to perform all of the duties of a homemaker as well."

For Jenny and many other female teleworkers, the sexual division of labour has not changed in their households. Paid work has been added to their responsibilities without any corresponding renegotiation of the family structure. The spatial logic within her home that relegates Jenny's paid work to a peripheral space both reflects the status quo within the family and perpetuates the subordination of her work by making it almost invisible to the other family members.

While almost all of the workers (92 percent) in the 1995 Canada-wide survey have a designated workspace at home, two-thirds have to share this space with other activities. Female home-based workers with children in particular select their workspace so that they can monitor family activities. Women are more likely than men to use a work area on the main floor that is part of another room. This corroborates data from the California study that demonstrates women are more likely to choose an area that allows them access to household activities when working. Male home-based workers are more likely to choose a work area that permits uninterrupted concentration. A study of Swedish teleworkers had similar results (Wikström, Lindén, and Michelson 1998). Women's homework spaces tended to be smaller and more spatially integrated into the dwelling than those of the men studied. A study of Mexican low-income homeworkers, though not teleworkers, also supports this observation (Miraftab 1996). Men tended to have separate, specialized home-based work areas. Women mixed their paid work areas with domestic space.

Ten years after the California study, the Vancouver study revealed comparable patterns. The women interviewed who were the primary caregivers in their families were more likely than the men to have work spaces that were shared with other uses, and women rarely allowed themselves the luxury of designated spaces. Karen lives in a large custom-built house on an island close to Vancouver. Her four-bedroom home, which she shares with her husband and three children, has a designated office on the lower level, spatially distinct from the rest of the house and with its own outside entrance, where her husband works. She works in a corner of her bedroom, which, while allowing her superb views, is not separated from the domestic functions of the home. Katrina also works in a shared space. A full-time teleworker based in Nanaimo on Vancouver Island, she does data processing in a nook in the kitchen. Though she has a three-bedroom home, she chose this working space because it was "a central area, close to kids, central to see kids." While it is in a very central location, she has impressed upon her children that "this is where I work."

Sen (1990) asserts that individual agency is influenced by a person's sense of obligation and perception of legitimate behaviour and that women have a strong sense of family identity that often negates their self-welfare within the family and maintains traditional inequalities. The women described above may be unwittingly conforming to the social norm that because they are women working at home, they are "not really working," and therefore their workspaces are not important. They may be habituated to inequality and unaware of the possibility for change.

Corresponding to these tendencies, male and female teleworkers' self-identities are reflected in the spaces where they work. Those who devalue their work, or more likely have their work devalued by others, choose spaces that diminish its importance. In contrast, those who have converted their living rooms to offices have a large investment in their identity as competent professionals. Their work has taken precedence over their social life and the time they spend with their families. Tom and Sylvia, the couple profiled in Chapter 3, both work at home but in different capacities. They have converted both their living room and a bedroom into offices. Tom's work for a government agency is done in the bedroom, while the consulting business he and his wife share is managed in the converted living room, further corroborating the tendency for women to select leftover spaces. They do their infrequent socializing in the family room.

People like Tom and Sylvia usually love their work or the rewards they get from it. In order to further their work goals they feel perfectly comfortable usurping home spaces for work-related functions. The home, for them, loses its special character as a retreat and becomes a utilitarian work site. In turn, the offices they have created at home resemble corporate offices with expensive desks, chairs, computer equipment, and subdued decorating.

Tom and Sylvia can separate their home and work spaces because they have a large suburban house. Others cannot. Brian, a consultant in electrical engineering systems in San Francisco, admits, "I don't have a home, I live in my office." He lives and works in a studio apartment. He sleeps in a Murphy bed that folds into the wall during the day. His kitchen is bare except for what he terms "office-like food supplies": a keg of wine, some packaged soups, and a few tins of beans. As a single person, Brian finds it hard to maintain a social life. He rarely entertains. When he is at home he is more than likely working.

Tim, the Vancouver creative director profiled in Chapter 5, feels his home is tainted by work. He works in his living room, which he shares with his wife and two small daughters: "I feel very invaded. There is no sense of sanctuary. Work is so integrated into home. Ninety percent of the time the computer is on and I find myself checking my e-mail when I should be attending to my kids. I feel very conflicted. It is almost like living in an office."

Joanne, an interior design consultant in San Francisco, also lives and works in a very small space, but because her apartment has two small living areas she has converted them to a separate workspace and living area. This separation of functions has allowed her psychological distance from her work.

Ann, the single mother of two teenage sons in Berkeley, California, works in even more confining circumstances in her small two-bedroom apartment. She works on her computer in her ninety-square-foot (8.4 square metre) bedroom or at the kitchen table. She sometimes has to set up the terminal and modem in her living room as well. Papers are spread out on her bed and floor. The telephone is in the living room with a long extension cord. She cannot work when her sons are at home and usually has to wait until they go to bed before she can start work again. She has no private space that is not work-related.

Few homeworkers can precisely define what constitutes their work environment. Though most have a designated workspace, they also use other parts of their homes to do their work. Work material is often spread over the living areas and the private bedroom spaces. Correspondingly, some homeworkers have home-related equipment in their offices, such as ironing boards, sewing machines, and baby cribs. Only a few have only office equipment and supplies in their home office.

Most homeworkers choose workspaces that are "left-over space": available, and not continually occupied by other people and activities. A few, however, choose their home because of their work requirements. George, the freelance radio producer and media worker from Oakland, California, chose to move from a house to a large studio with a loft that he uses for both his living and working space. He wanted to be closer to his work and have fewer distractions to keep him from his work. Sally, the medical transcriptionist from Hercules, California, moved to a new house with her family specifically because it had a room that met her specifications for a workspace. The room was on the main floor, had plenty of light, a view of her backyard and a bird feeder, and was adjacent to the kitchen, bathroom, and laundry.

What homeworkers generally like about a workspace, regardless of where they work in the home, is that it is much quieter than an office and has more comfortable surroundings. Especially important to homeworkers was the control they had over planning their spaces. Several homeworkers discussed in detail their choice of paint, window coverings, and furniture for their offices. They were concerned with creating an identifiable office environment that was more conducive to their needs than a corporate office. They wanted it to be professional, but serene. There was also concern about making their office comfortable and more "homelike." Good, comfortable chairs were selected. Most people have plants in their offices; many have family pictures.

Views, vistas, access to the outdoors, and natural light were important criteria in homeworkers' satisfaction with their workspaces. As one homeworker emphasizes, "I could be very isolated and cut-off if I didn't have this vista [from my workspace]. I can see downtown from here, I can see what my neighbours are doing, I can see what the weather is doing." The lack of these amenities was especially felt by those who didn't have them. Their workspaces felt cramped and bleak.

While most people like some aspects of their home workspaces, many have problems in their use. Most people have trouble controlling the amount of paperwork they generate. Jane, with several different businesses, works in a converted garage on her word processing and uses her dining room table and living room to store material for her other projects. Stacks of paper prevent her from using her table. She admits that she is embarrassed to have people over because of the mess and chaos in which she lives. She lives alone in a townhouse in Marin County but still finds she does not have enough space for her work.

Mark, a tax consultant in Vancouver, has invaded almost every room in his house with his work. His wife, Rachel, estimates that three-quarters of their renovated bungalow, in a quiet residential Vancouver neighbourhood, has work-related activities. Mark's main office is the converted living room, and he holds client meetings in the dining room, where they no longer eat since there are stacks of work materials everywhere. As well, he stores files in the basement and in the guest bedroom, which can no longer be used for visitors. Often, Rachel and their ten-year-old daughter come home to find clients sitting in their family room, which adjoins the kitchen. Even Rachel's sanctuary, a small alcove adjoining their second-floor bedroom, has been invaded. Because Mark's computer does not have the capacity to load a new tax program, he uses her computer in the alcove office. Rachel recognizes that if he were working elsewhere she would not see him for months at a time, but she has concluded that she needs to find an office outside of the home that is her own.

Even those who have been able to control the spread of their work through the home find that they don't have enough room in their workspaces for filing, desk, and shelf space. Inappropriate space was a prevalent complaint for homeworkers, most of whom had little choice of workspace. Often people's choice of a workspace was dictated by their need to find one designated room in their home that they could write off as a workspace for income tax purposes. Besides size, home spaces that are converted to workspaces are not adapted to office requirements. Homes lack proper wiring for computer use, precipitating electrical failures and the inability to use household appliances while the computer is running. Older homes, especially, lack adequate temperature control systems, making the workplace either too cold or too hot.

Currently, homeworkers are living and working in spaces traditionally meant for home activities. The activities of telework and home-based employment often dominate the home environment both spatially and temporally, changing the nature of the home and home life. Rather than integrating home and work life, home-based work is causing conflicts in the use of the home and in the way that the homeworker interacts with the rest of his or her household and community. An understanding of these conflicts has implications for the planning and design of homes.

What becomes apparent after interviewing people who work at home is that many homes are unsuitable workplaces as presently planned. There is no clear division between home and work functions, creating time and space conflicts that interfere with a household's functioning. For homeworkers, the home is not a refuge, but a utilitarian space. Nevertheless, these perceptions also depend on the type of housing unit, neighbourhood, and access to neighbourhood resources.

Housing Profile of Home-Based Workers

Workers in all three samples predominately live in and own a single-family detached house of recent vintage. In the Canada-wide study, the majority (75 percent) live in single-family detached housing in urban or suburban communities. Their homes, most typically, have three or more bedrooms and a floor area of over 1,600 square feet (149 square metres). Their mortgage or rent is usually over $635 per month. They have, on average, lived in their present home less than six years. Though independent contractors and self-employed entrepreneurs are the groups that spend the most time at home, they have, on average, the smallest units.

The use of the home is influenced by its appropriateness for various home-based work activities. The home is used most frequently to make or receive business calls, to do administrative work, to provide professional and other services, and as a mailing address. It is less used for customer meetings, for manufacturing, or for processing goods and services. Interestingly, both men and women use the home for work in similar ways. The difference is between teleworkers and self-employed homeworkers. While both groups use it for telephone calls, administration, and professional and other services, self-employed homeworkers are more likely to use the home as a mailing address, to store goods and equipment, to have client/customer meetings, and to manufacture goods.

Close to two-thirds of teleworkers indicated that the primary workspace in their homes was chosen because it was easily convertible for work activities and in the right location to minimize impact on the rest of their home. Employed teleworkers tend to use more common areas (such as main floors) and share areas with other functions (such as bedrooms). Self-employed homeworkers tend to have workspaces located far from other activities (such as basements). Home-based workers who live in single-family detached houses are more likely to have designated workspaces than those who live in higher-density housing. This is probably related to the size and number of bedrooms in the dwelling. Female home-based workers are less likely than men to have a designated workspace. One-fifth of the respondents had converted a living room into a workspace, and one-tenth – all single men – used their whole home as an office. These men describe their homes as an office that they live in, rather than a home where they work. Every area of their "office home" has a work-related function attached to it.

While most home-based workers have a dedicated workspace, two-thirds of the sample reported using other areas of the home to do work. Many read business reports in the living room and bedroom, write on the dining room table, and conduct business telephone conversations almost anywhere. Those who have an office/study or workshop use it most often for work, but they often work in rooms where other activities occur, such as eating, socializing, and sleeping. The rooms with the most overlap of activities are the kitchen/dining room, the living room, and the bedroom. A sizeable minority (varying from 30 percent to 45 percent among the three studies) used workspaces that were shared with other activities.

Other studies corroborate these findings. Ahrentzen's 1987 survey of 111 homeworkers found that 70 percent had an exclusive workspace, while 10 percent shared a workspace with occasional activities and 20 percent shared with daily activities. A survey of 373 readers of *Personal Computing* found that 61 percent had an exclusive office space, 9 percent used the dining or living room, 7 percent the family room, 5 percent the den, and 4 percent the kitchen (Antonoff 1985). Interviews by Christensen (1986) with thirteen corporate-employed homeworkers, all women, found that seven had exclusive workspaces, four shared their workspaces with their children, and two shared their workspaces with their husbands.

Only five respondents (1 percent) in the Canadian survey have a designated workspace/office in a detached building adjacent to their home. All five are home-based business operators who own their own homes. They find that their workspace arrangement addresses their need to separate their work from their household activities but still allows close contact. Twenty-six other respondents also work in a detached building, but it is also used for other activities. Most of the respondents who work in a detached building are in manufacturing/processing, with the rest in retail trade/product sales, personal and professional services, agriculture, construction/trades, and wholesale trade.

Some home-based workers had their houses built with a home workspace as a priority in the design and layout, while others bought a home because it had a space specifically designated for work. Female home-based workers particularly selected their workspaces so that they could be close to family activities that need monitoring. Other reasons include the need for privacy from the rest of the household and easy access to other areas of the home such as the kitchen.

Problems with Working at Home
Some of the major problems with working at home include lack of storage for materials and products, and intrusions from family, neighbours, and friends. Other complaints include workspaces that are too small, unsuitable layout for working, poor lighting, electrical wiring, and ventilation, inadequate number of telephone lines, and noise disturbances from outside the workspace. Home-based workers with the smallest floor areas in their

homes generally had the most problems with their at-home workspaces. They especially lacked adequate storage space and found their workspaces too small.

Fifteen percent find that some of their work-related activities were incompatible with their home environment. These activities include frequent work-related telephone calls and small-scale manufacturing that produced noxious vapours. Other problems are inconvenient access to workspace through a living space, a home layout unsuitable for home-based work, and an unprofessional workspace inappropriate for receiving clients. Problems outside the home include lack of space for loading/unloading/delivery of materials/finished products, inadequate employee/visitor/client parking, storage of hazardous work-related materials, opposition from neighbours to work activities, and complaints from municipal agencies regarding zoning infractions, incompatible uses, and so on.

Female at-home workers recognize the importance of an organized and maintained workspace with clear boundaries. One woman described the problem succinctly: "Home-based work takes over the house." Physical changes that they have incorporated include locking doors to their workspace, better lighting, and electrical extensions. Others have added more storage. Several find their present working arrangements totally unsatisfactory and are planning to move to a larger home or build a workshop in a separate building.

Male home-based workers have similar problems, such as messy, disorganized workspaces and difficulty keeping their families out of their workspaces when they are working. They have resolved some of these problems by building more storage, adding a business-only telephone line, and improving their lighting and technical equipment. Several are planning major renovations of their home to make it more suitable for work.

Since they have started working at home, the respondents find that, except for garbage, their households have not generated more noise, sewage, traffic, odours, or chemical waste. This is because few people are engaged in work of a hazardous nature. Nevertheless, a few at-home workers encountered problems with the licensing requirements of their municipality. Some were refused a business licence, and after negotiation now operate under stringent rules, such as no advertising, no posted hours, and no signage. Others are not allowed to have customers in their homes.

The home-based sample in the Canada-wide study were asked whether they had moved or contemplated moving because of their home-based work. One-third of those who responded to this question have already relocated or want a larger home with a more appropriate layout or more nearby amenities. They tend to want to relocate in the same neighbourhood. Others have moved from condominiums in the city to a country home, or moved back to their birthplace from the city to start a small home-based business.

While many in the sample contemplate moving due to the requirements of work, a few home-based workers believe that their present home and its location is ideal. As one comments, "Where I live is in an excellent location so any change with my business in mind would be to a larger residence extremely close to where I live and work." Another comment was, "As only administration work is done in my home, where we live is fine for now. Our location is superb." Some have already custom-built or renovated their homes to make them suitable for home-based work.

Others would not consider moving their home but might consider moving their work out of their home to a business district if it got too big or their privacy decreased. One person made a representative remark: "The location of the home would not be based on the type of business going on within it, but whether or not it was conducive to raising a family."

Those who are contemplating moving recognize the problems of making a home fit the requirements of work. As one respondent notes, "We are indeed moving out due to lack of space. It is difficult to disconnect from work if the telephone rings during dinner time." Another person concurs, "We would like to buy a three-bedroom house with the purpose of distinctly separating the office from the two bedrooms." Several people are concerned about the inappropriateness of their home for clients, and one noted, "I need a more convenient area for clients. I need a show room but the zoning does not allow it." Others want to stay in their homes and adapt them to their needs. One female home-based worker commented, "There is a difficulty in separating work and family life. Sometimes I think of building a cottage on my property to get away from work and chores." Another person agrees, "I have not considered moving, but have given serious thought to building a new work site within my existing home such as adding an office in a future basement development." In the diversity of responses of those contemplating a move, there appear to be no simple or clear shared reasons.

The respondents also recognize that working at home provides opportunities for flexibility in location. Several people commented that since their business revolves around "phone/fax/modem/computer use" and will do so even more in the future, they don't anticipate needing a lot of space for their work nor having to move from their present home because it is unsuitable. As one person stated, "Most of my work is done by phone, fax and computer so the location is no longer that important." Another teleworker who constantly uses electronic communications commented, "If I move I must be in an area that long distance charges [don't] apply to my modem connection." This remark has implications for the preferred location of homes, as teleworkers and other frequent users of telecommunications technologies would want to locate where they had access to appropriate and affordable telecommunications connections, such as fibre optic cable. Several respondents who live outside of major metropolitan areas voiced a similar concern. As noted by one, "The availability of clean phone lines and direct access to datapac and Internet computer services is first priority (e.g., it costs $30.00 per hour to access computer lines from where I live in a rural location)." Because of their small populations, rural areas have the least advanced telephone services and telecommunications capabilities.

Attachments to the Home

Though many homeworkers are living in homes inadequate for the mixing of home and work life, the home itself takes on great significance. They feel a great attachment to their homes that they did not feel before they started working in them. No office can come close to the pleasurable experience of working in one's own space, they say. There was also a consensus that homeworkers felt responsible for the upkeep of their homes, which they experience during all hours of the day and in all weathers. Interestingly, while women generally expressed strong feelings of attachment to their homes whether or not they worked there, men were generally surprised to discover what strong attachments they developed to their homes, because those feelings were not prevalent before they started working in them. One male homeworker explains: "I have got to know my home a lot better functionally. It's more fully part of my life. Before, it was just a dormitory and cafeteria."

A few homeworkers liked their home environments so much that these feelings became a significant factor in wanting to work at home. Sally, the medical transcriptionist portrayed in Chapter 3, felt that her home was a "money maker" for her. Because she was so happy in her environment, she was more productive. Others, because they didn't have enough time to spend on maintaining their homes, felt the home was less a personal expression of themselves and more a functional setting for activities.

The strong feelings that home-based workers have about their homes are in not all cases positive. Often, the home is perceived as claustrophobic, confining, and isolating. Bob, the San Francisco teleworker profiled in Chapter 3 who has worked at home for five years as a freelance writer and editor, eloquently describes his feelings:

> The merging of work and personal life is a crazy set-up. I feel very resentful about it ... My whole life is one big lump – doing everything at the same time. I feel like I am in a minimum security prison. I am doing productive work in a prison with no place really to go. My whole home has become my office. Every room has paper in it. I can never retreat. My neighbourhood even reminds me of work, as I do some of my work at local cafés. I have to go out of town to get out of the work mode.

For part-time homeworkers and office workers, the home does not have such strong connotations. Their feelings about home are generally not articulated in terms of strong attachment to a physical environment or location, but rather "where my family is" or "an expression of my wife's and my relationship." For single people, home is a retreat from the world, and a place to sleep and occasionally eat. Office workers rarely get to enjoy their homes except on the weekends. As one married male office worker from the California study comments, "My home is a neutral ground. I don't have real strong feelings about it. I can be comfortable almost anywhere. But when I am at home things that I have to do around the house do nag at me."

The home becomes a much more important locale for homeworkers than office workers. It is perceived as a far better work environment than a corporate office. Nevertheless, because of its importance, it can also become confining, as there is no respite from it. Homeworkers have the double-edged problem of lack of community and lack of distance from their work. Most homeworkers no longer feel totally comfortable at home because they have no separation from their work.

The Home of the Near Future

> Imagine yourself in a house that has a brain, a house you can talk to, a house where every room adjusts to your changing moods, a house that is also a servant, counsellor, and friend to every member of your family ... the application of computer technology to architecture is transforming that idea into reality (Mason, Jennings, and Evans 1984, 17).

Two recurring themes that have historically pervaded discussions of the home of the future are still prevalent today: the home as an autonomous entity and the home as a socializing agent of society. The ideal, represented by the self-contained single-family house, has always been counterbalanced by the dependency of the home on municipal and commercial services. Futurist speculations on the home of the future fail to understand the recurring tensions in American society between independence and individualism, and the desire for connection and caring, that are physically manifested in the home.

Twentieth-century theorists and designers have offered numerous visions of "the home of tomorrow" (Corn and Horrigan 1984). For them, the home is no longer merely a physical structure that people occupy, but a machine that, through embedded electronic control systems, can respond to changing personal needs. Technology has made it possible, these designers feel, to realize the "home as womb." These visions, imbued with the belief that home life should be made more efficient, derive from changes that occurred in the industrial and urban landscape of the late nineteenth century. In response to industrialization, it was thought that a proper home, appropriately and efficiently run, could prevent societal moral decline. Catherine

Beecher and her sister Harriet Beecher Stowe ([1869] 1975) instructed their readers on how to manage and maintain a gracious home in a practical and moral fashion. Others developed a more radical analysis, suggesting that industrialization could be used to develop technology that could socialize housework and child care, freeing these tasks from the constraints of the home (Bellamy [1888] 1967; Perkins Gilman 1919).

Technology plays a key role in the literature on visionary homes. In the futurists' writing, technological tools eliminate the drudgery of housework and allow for the real individualism of "man and his family" (Fuller and Marks 1973). Integral to this vision of a decentralized future is the technology that allows people to perform their jobs at home while connected to distant employers. The Dymaxion House proposed by Buckminster Fuller in 1927 exemplifies the belief that the home should be more responsive to machine-age imperatives such as mass production, mass communication, decentralization, and mobility. The house had a "get on with life" room, equipped with a typewriter, calculator, telephone, dictaphone, television, radio, phonograph, and mimeograph machine, all in one factory-assembled unit.

Newer versions of the home of the future have been proposed with the same basic philosophy but more sophisticated gadgetry. "Xanadu," a futuristic foam house, and "Sculptec – The Intelligent House" are equipped with technologies that give the house of the future a brain: a central computer that responds to human speech, controlling and monitoring the home's environment and appliances, and communicating with other computers around the world (Mason, Jennings, and Evans 1982, 1984). Integral to this vision is the belief that new technologies will free people and their children to live a home-centred life, with all work, education, and leisure activities conducted at home. This concept of a home where so many functions can be performed reflects an anti-city bias on the part of the visionaries, who assume that having to deal as little as possible with urban life is inherently desirable.

Designers and developers are currently planning the electronic home or "smart house," which combines various computer-enhanced security, telecommunications, energy-monitoring, entertainment, and climate-control systems to create a self-sufficient environment (Smith 1988). Various computer-activated sensors and control systems regulate and modify environmental conditions within these homes and in relation to the outside world. Designers of these environments view electronic technology as a liberating force that will free people from household drudgery, make work more creative, allow instant gratification through entertainment, and provide protection from unwanted intruders. The household they envision is one in which individuals can creatively manipulate technology for their own ends and privacy is attained by retreating to individual "technological pods."

This new form of home has the potential to allow work, leisure, and educational activities to occur in the home and permits services to be managed from the home. Through computer and modem linkages to company databases and coworkers, workers can turn their homes into "electronic cottages." Computer networks allow the exchange of current, specialized information between scholars, scientists, and hobbyists around the world. Computer databases provide opportunities for electronic information gathering, education and home-based banking, shopping and ticket ordering. Telecommuting, teleschooling, and teleshopping can now occur in the home, making the home not just the centre of a household's universe, but the universe itself.

Besides the outside activities and services that can be provided in the home, the electronic systems in the smart house make it possible to connect home devices and appliances in a network. This network allows the home owner to program these devices and appliances for maximum comfort and convenience. The home is increasingly being seen as a "communication hub" equipped with home computers, videocassette recorders, and cable television hook-ups, in addition to the telephone and television. Through highly specialized telecommunications equipment, homes may be transformed into "electronic hearths" with entertainment provided by new technologies that will augment many of the leisure-time activities already associated with the home.

Design proposals for homes of the future emphasize creating rooms for togetherness based on common telecommunications use. The home as a place for entertaining friends is de-emphasized. Rooms that are not individual private spaces become multi-purpose rooms. The living room, traditionally seen as a "don't touch room," is replaced by an all-purpose dining, socializing, and relaxing area. In one design this "us room" is equipped with a fireplace, a large-screen, built-in television, a family computer space, and eating and seating areas (Figure 6.1; Johnson 1990). Radiating from this central space are the other, more private, rooms. The philosophy behind this design is that the "us room" will become the hub of activity; bedrooms are small, merely places to sleep.

The electronic home as a phenomenon has garnered considerable interest because of speculation it will transform the social and economic life of society. Futurist writers (Toffler 1980; Mason, Jennings, and Evans 1982) envision a home-centred life where all social activities and transactions are electronically mediated. They perceive information and telecommunications technologies as the basis for improvements in human fulfilment and welfare. Home-based activities, say these futurists, will foster community stability, and family and friendship ties. While these visions have lofty goals, they fail to recognize the evolution of the family into smaller and more fragmented units. Families rarely eat together and hardly ever share

Figure 6.1
_____

Home for togetherness based on telecommunications use

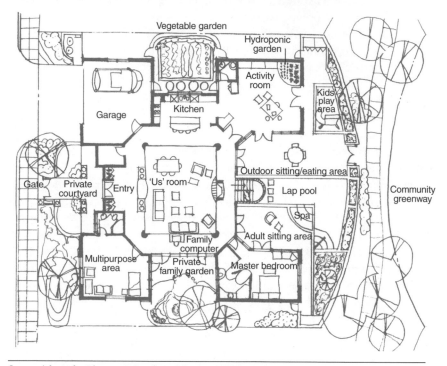

Source: Adapted with permission from Johnson (1990).
_____

common home activities. An alternative vision has the home being used as a series of personal work and entertainment spaces serving the individual needs of family members. For example, a New York City apartment has been designed as a series of studio-like spaces, incorporating videocassette recorders, compact disk players, and computers in each room (Figure 6.2; Slesin 1986).

While there is a growing recognition of the need to incorporate working at home into the plans for new housing, there is very little understanding of what that entails. Many books and popular magazines suggest layouts, equipment, and furnishings for the home office (Edwards and Edwards 1985; Atkinson 1985; Scott 1985). Merchandisers are now advertising furnishings solely intended for the home office. Though these publications offer information and advice, especially on electrical wiring and ergonomically designed furniture, little is known about teleworkers' actual preferences for home workspaces.

Moreover, zoning ordinances often forbid or restrict home occupations, making it difficult to incorporate plans for home offices in new developments.

Figure 6.2

Home as a series of studio-like spaces

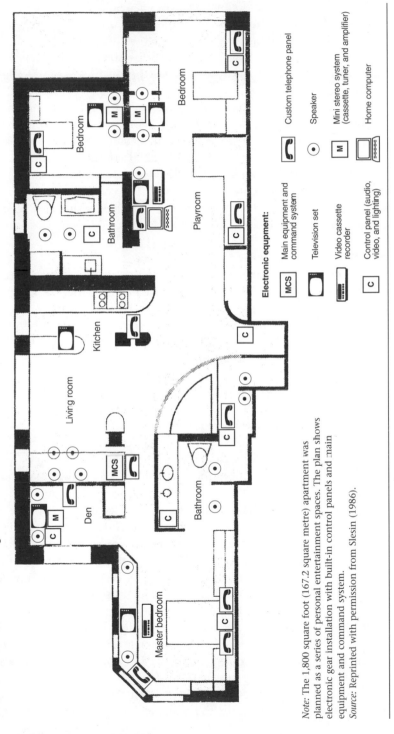

Note: The 1,800 square foot (167.2 square metre) apartment was planned as a series of personal entertainment spaces. The plan shows electronic gear installation with built-in control panels and main equipment and command system.
Source: Reprinted with permission from Slesin (1986).

However, the belief that occupational uses generally do not belong in residential zones arose in response to the health hazards of industrialization, concern about pollution, and the destruction of neighbourhoods. These zoning practices are no longer relevant to a postindustrial society with an emphasis on service- and knowledge-based industries, and are gradually changing.

A few developers have made workspaces an integral part of their residential designs. Artists' live/work spaces are the most visible examples in the urban landscape. These include spaces in which the living and working areas merge, such as artists' lofts, as well as spaces that separate the studio space and the living quarters. These live/work spaces, however, are generally not intended for the use of a family. The complexity of combining working activities and family life in the same milieu has not been fully thought out by developers.

The "Ideal" Environment for Teleworkers

The research outlined in this and previous chapters has found that teleworkers do not live and work as envisioned by writers in the popular press. In addition, the preferences of teleworkers regarding their "ideal" environments are vastly different from what designers are currently contemplating. Crucially, there is no one ideal environment for homework. The ideal home workspace varies according to the needs, occupation, and sex of the homeworker. Nevertheless, several patterns emerge regarding the most desirable spaces. Professional identity is a critical factor. Those homeworkers whose identities are strongly connected to their work prefer spaces that reflect as professional an image as possible. They prefer separate entrances, and in some cases a separate structure, for their workspaces. Those whose primary concern is the care of their family do not want their workspaces to be separate from their home, but they do want to work in a separate room.

Home-based workers have a definite idea of a suitable workspace. It includes natural lighting and ventilation, visual and acoustical privacy, and adequate storage and electrical amperage and outlets. Almost all of the homeworkers studied want a separate room for their workplace. This aspect of the ideal workspace is critical. Most felt that they could not function as paid workers in environments that are used for other activities, even though many are presently doing it. Male home-based workers tend to want their workspace to be as separate as possible from the home space. As one male homeworker firmly believes, "The more discrete and distinct the workspace is from the house the better. Keep home life and work life separate. I want to be far away from places, like the den and kitchen, that make me draw away from work ... where other people could be." Female homeworkers want separation but with close proximity to home activities. For example, most male home-based workers (two-thirds in the California study) want a workspace that cannot be seen from other rooms; the response was more mixed from female home-based workers. Women with children were also less concerned about privacy and didn't consider environmental features such as views and private outdoor spaces as important as their male counterparts did.

While most of the home-based workers consider a view from their workspaces important in relieving the tedium of staring at a computer screen and in mitigating their feelings of isolation, male homeworkers and female homeworkers without children generally want a view out to the street. In contrast, female homeworkers with children do not consider a street view as important; if they had the option, they would prefer a view of nature in their backyard. This preference can be attributed to the fact that many of the homeworking mothers live in neighbourhoods where few activities occur during the day in the street. Consequently, they prefer the view of a serene garden to an empty vista. Most male home-based workers want a private outdoor space attached to their workspaces; only half of the female homeworkers consider this desirable, and most do not consider it a high priority for their ideal environment. These women would prefer to leave the home entirely if they wanted to relax outdoors.

Working only part time at home also affects opinions about the ideal work environment. In the California study, both male and female part-time homeworkers are the least likely to want a separation between their home and workspace. They are not as concerned about privacy and environmental features as the full-time workers. However, in the Canadian survey, teleworkers (most of whom work part time at home) have similar priorities to the rest of the sample for their ideal home workspace: a separate room for work, natural lighting, proper ventilation, and acoustical privacy are most important. Teleworkers consider adequate power outlets to be a somewhat higher priority than the rest of the sample does. Both groups consider adequate storage space very important.

Most homeworkers want a larger workspace than they have now, indicating that a space at least twice the present size would be ideal. As the majority of homeworkers are working in small converted bedrooms of 100 square feet (9.3 square metres) or less, the ideal workspace would be at least 200 square feet (18.6 square metres). This finding corroborates Ahrentzen's 1987 study, which calculated that an ideal median size for a workspace would be 240 square feet (22.3 square metres). Rather than one large space, the ideal is several smaller spaces for reading, working, and storage of materials that would be planned to encourage an efficient work flow.

An ideal home workspace would also have significant environmental comforts to alleviate fatigue. The furniture and equipment would be ergonomically designed and adjustable. The office arrangement would be planned to prevent glare on the computer screen. The workspace would make the best use of natural light, and have effective artificial light and task lighting. Colours and textures would be used to create a soothing atmosphere. There would be enough space to reach things easily and move a chair around.

Many home-based workers recognize that having their work located in their home allows them to choose a home and its location based on a set of priorities that allow for the integration of both home and work. Some people were very specific about what environment they would like and one commented, "I would want a home with a layout all on one floor, with the office near the entrance and large enough to conduct all business activities."

Opinions vary among homeworkers about whether to live closer to or further away from a city. Some would like "to move far enough out of the city to be away from [the] city but close enough to benefit from the city." Others would "love to have a home-based business on an acreage in a rural setting," while others prefer "not too far from the town centre but above all close to services and bus." Some would prefer a detached building for work, such as one person envisions "in a rural area with space for a separate workshop and distance from neighbours because of noise from woodworking machines." In general, at-home workers recognized the importance of accessibility to services.

A Live/Work Typology

A typology of live/work space configurations has been developed to reflect the range of homeworkers' experiences. Four distinct types of spatial relationships have been identified – "Work dominates," "Live/work blended," Live/work separated," and "Work shared" – corresponding to the degree of work penetration in the home sphere (Figure 6.3). These should not be taken as ideal types. Instead, they describe the range of spatial possibilities for the live/work experience. As well, the preferences of the homeworkers vary individually and are only tendencies, which vary according to sex, age, household type, occupation, and financial resources. In addition, preferences may change based on the experience of working at home. For example, male homeworkers with children might initially choose to work at home because they want more involvement in the domestic sphere, but once they start working at home might find that they need more separation. Conversely, some men with children might initially choose to work in physically separate home offices but then choose to move their offices closer to domestic activities.

"Work dominates" describes a live/work situation in which work dominates domestic life, leaving little space for non-work activities. "Live/work blended" is a spatial relationship in which the home environment is blended with work but there are varying degrees of separation between these two spheres. "Live/work separated" is a home/work relationship where home and work activities are physically separated, but are in the same structure or

Figure 6.3

The live/work space typology

Live/work type

1 Work dominates
Work dominates domestic life leaving little space for non-work activities. This arrangement is primarily inhabited by solos or couples without children.

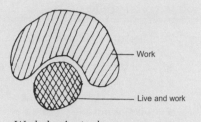

Work dominates home space

Examples

A Home in studio workspace
A studio apartment in which work is the main function of the space

B Live/work arrangement in a loft studio
An artist's loft live/work studio. The one large space is not subdivided except for a small bed area in a loft above the main space.

C Separation of working and relaxing
Living and working share one space, but there is a separate, spatially defined area for relaxing and socializing.

▶

◄   *Figure 6.3*

Live/work type

2 Live/work blended
Part of the home environment is blended with work but there are varying degrees of separation between these two spheres. There is no buffer between work and domestic spheres. This arrangement is the tendency of home-based workers with children who want close proximity to their children's activities.

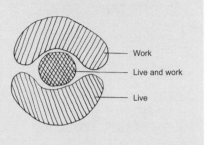

Work and home blended

Examples

A Office in home in close proximity to domestic activities
The workspace is integrated in the home, close to domestic activities but physically separate.

B Office in home with possibility of separate entrance
A home office in which the workspace is separate from the other areas of the home, off the main entrance. As the dotted outline of the door in the workspace indicates, it may also be planned to have its own street entrance. In this scheme, the living room space is minimal, while the bedrooms are generous to accommodate personal activities.

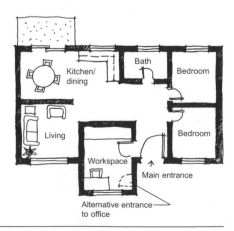

▶

◀ *Figure 6.3*

Live/work type

3 Live/work separated
Home and work activities are physically separated and have different entrances, but are in the same structure or on the same housing lot. This arrangement is the tendency of homeworkers with children but who also have clients visiting them regularly and want a high degree of separation from domestic activities.

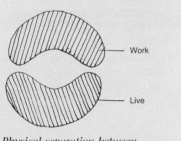

Physical separation between working and living

Examples

A Vertical separation of home and work areas
The home office is on the ground level of a two-storey house with its own entrance off the street. The diagram shows a section through a home that illustrates this separation. The office has a semi-public meeting area for clients and a private office for the homeworker. In this situation, the home-worker gets a view of both the street and the private backyard. A door leads from the office area to stairs that go up to the living area on the first level. The private entrance is physically separated from the office entrance and also leads to stairs to the first level.

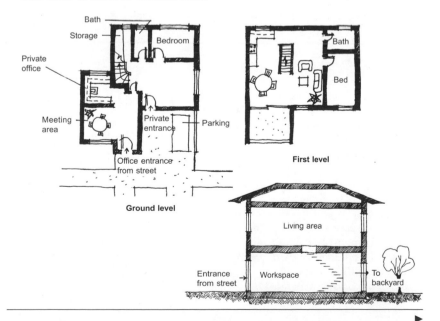

▶

◀   *Figure 6.3*

B Workspace in separate structure

The workspace is in a separate structure, such as a converted cottage or garage, from the main house. It also has a separate entrance off the street.

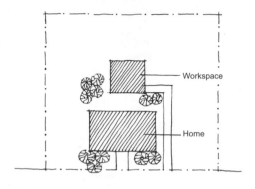

Live/work type

4 Work shared

The workspace is physically separated and shared by a group of homes. Though this alternative does address the needs for both community and privacy, it would be a radical departure from existing housing developments in North America and would probably require special zoning considerations.

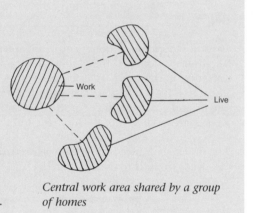

Central work area shared by a group of homes

Examples

A Clustering of homes, work, and recreation

A housing development with a common work centre adjacent to a social/recreation centre. The work centre is located in the heart of the development, is shared only by residents of that development, and the rent for the workspace is incorporated into the monthly maintenance costs. The individual workspaces vary according to specific requirements, are separated by partitions, and can be locked.

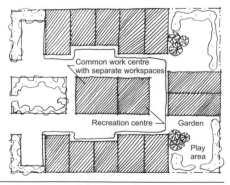

▶

◄ *Figure 6.3*

B Work separate from homes

A housing development has a common work centre that is separate from the housing component. Space in this work centre could be rented to other people who are not living in this development. This work centre could be situated as the "public face" of the development.

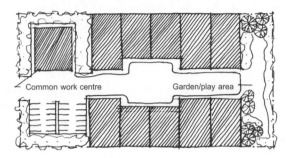

C Clustering of work in neighbourhood work centre

A work centre is located in a neighbourhood to attract a wide number of potential workers who could walk to work. The work centre would have support services for the small entrepreneur as well as a child care centre, a café, a convenience store, etcetera.

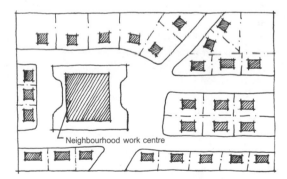

on the same housing lot. "Work shared" describes a relationship where the workspace is physically separated and shared by a group of homes. The typology was developed to speculate on the range of possibilities. Some of these solutions may alleviate certain negative aspects of working at home, but on their own they cannot solve the social and psychological problems of home-based workers. This would require fundamental changes in how the domestic and work spheres are viewed and accommodated in North American society.

As currently planned, most homes need to be adapted for homework. However, the research on work environments has primarily been carried out in offices, and little is known about the environmental aspects of using the home as a workplace. The most appropriate design solution for a particular homeworker depends on his or her particular work, household structure, housing situation, and financial capabilities (among other criteria). Figure 6.4 diagrams the considerations that address the needs of the diverse home-based work population.

Before people begin to work at home they must evaluate their needs and the resources they have available. Too often a home workspace is selected with very little thought, leading to awkward spatial arrangements that affect the functioning of the home. The specifications for an appropriate home workspace depend on individuals' work behaviour: what tasks they do, what equipment they need to accomplish the tasks, and how they screen people and conditions that tend to interrupt the flow of work. Physical work problems such as eyestrain, headaches, and backaches, and emotional problems such as feeling fragmented, hassled, angry, depressed, and bored should be identified. Particular environmental and social criteria such as the need for a view out from the workspace and the image that the home-based worker

Figure 6.4

Considerations in designing homes with combined living and working spaces

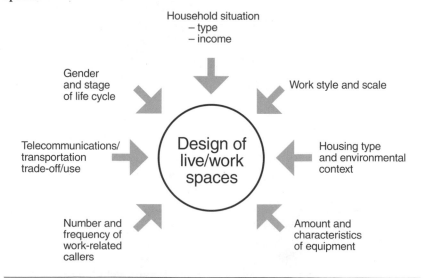

Source: Adapted from Cullen et al. (1989), Appendix 3.2-11.

wants to create in the workspace should also be clarified. If the home-based worker is a teleworker, the employer can prepare a set of guidelines regarding the servicing and use of equipment, and the layout of the home office.

Of crucial importance for planning workspaces in the home is the concept of privacy. The need for psychological, acoustical, and visual privacy varies among individuals and cultures. While privacy refers to the ability to carry on activities free from intervention or observation, isolation is created when a person is denied social opportunities when desired. The difference between privacy and isolation depends on the degree of control that a homeworker has over his or her environment and the homeworker's freedom to relinquish privacy at will. Home-based workers have difficulty separating themselves psychologically from work if there are not clear boundaries between their home and work spaces. In turn, household members can easily interrupt homeworkers if they can be seen and heard from spaces used by the rest of the household. Privacy can be achieved in a number of ways – through physical barriers, spatial organization, scheduling, and codes of behaviour. These include defining the workspace by an arrangement of furniture or screens; establishing social rules regarding the use of the space; and scheduling times for the use of the space.

Once home-based workers' needs are understood within the context of their households, the various elements of the home should be evaluated to ascertain how these needs can be accommodated. Can the at-home worker's needs be accommodated in the existing home or do adaptations have to be made? Is there enough separation between the chosen workspace and the other areas of the home? If the home cannot be adapted or the needs of the homeworker cannot be accommodated at home, are there alternative workspaces close by? These questions (among others) need to be addressed in order to ensure a comfortable fit between the user and his or her environment.

Consequences for the Use of the Home

The use of the home as a work environment affects the role of the home. While the futurists have predicted that home life will be strengthened by homework, the findings presented here have identified major conflicts in the integration of home and work life. Empirical research has indicated that for homeworkers, the home loses its nature as a place of refuge and acquires more of a confining or isolating sense. This experience is similar to that of unpaid domestic workers. The research has found that while working at home has distinct advantages for the employed teleworker, many self-employed and independent contractors are overworked and working in inadequate conditions.

Working at home generates new activity patterns that affect the household's structure and the internal organization of the home. These activities are part of larger societal processes, which are locally situated in the home. The built environment influences the development of social images by either supporting or hindering expected roles, identification, and status. Role settings, or "stages" (Goffman 1959), are territories that define a situation and are identified with, and under the control of, individuals or groups. These settings are a means of achieving privacy and identity by controlling information and access. When domestic and work activities overlap, social roles merge. Conflicts between family and work roles have been identified as a major problem for home-based workers, and these conflicts have been observed to have both spatial and temporal manifestations. Since homes are not designed for work, work has to be done in spaces designated for other uses. Clients visiting the home contribute to conflicts, since the normally private home must assume a professional identification. When these conflicts cannot be resolved, the merging of roles and the merging of spaces disrupts the activities associated with the home. To minimize conflicts, spatial, temporal, behavioural, and psychological boundaries can be constructed. The disruption of the home affects the delineation of the private and public realms of the home.

The critique of Goffman offered in Chapter 1 argued that an improved work culture and workspace does not necessarily liberate an alienated worker, as Goffman would maintain. The daily life patterns and workspaces of teleworkers bear this out. While most prefer homework to working in an office, the spatial and temporal conflicts and lack of clear boundaries found in the workspaces of teleworkers, especially women, result in an ambiguous working identity, further adding to the tensions in trying to maintain both domestic and work life.

The live/work trend has major implications for urban design and community infrastructures. Will working at home precipitate the decentralization of resources? Will the use of neighbourhoods be intensified? In what ways do mixed-use inner-city neighbourhoods and segregated-use suburban communities support or hinder the activities of homeworkers? These questions will also be addressed in the next chapter.

Though designing a home workspace suitable for homeworkers is clearly important, the real challenge in planning home workspaces is to find ways of incorporating opportunities for community and privacy within the same setting. Now that social connections can be made, and information electronically received in the home to a greater extent, new definitions of privacy need to be examined. The successful integration of work into the home environment will require planning that incorporates a clear separation between home and work activities within the same structure. It will also require a reconceptualization of "home" both in the greater society and in the individual household.

Meanings of Home

The fundamental question of the meaning of "home" is a recurring unresolved theme in environment-behaviour research. Rapoport asserts that meaning in settings refers to "nonverbal communication from the environment to people" (1982, 178). In turn, meaning locates people in social space, thereby influencing communication. Though meanings are inherent in people, not objects, environment-behaviour researchers support the notion that things do elicit meanings that can be decoded by users. Because of the complexity of defining "meaning," it has often been used interchangeably with "role" in asserting that meaning communicates the context of a social setting (i.e., who should interact with whom, when, and under what conditions) as well as providing information about status, lifestyle, ethnicity, and other variables.

While this definition may describe a certain type of person-environment relationship, it neglects the role of socioeconomic forces in people's interpretations. Despres (1991), in a review of the literature on the meaning of home, differentiates between the microsociological interpretations that suggest that the meaning of home is the result of biological, psychological, sociopsychological, phenomenological or developmental processes, and the macrosociological theories that suggest that the meaning of home is the result of historical and contemporary ideological discourses and political-economic institutions, as well as the formal structural properties of the home. She argues for an integrative approach that investigates environmental meaning from the perspective of both societal forces and the individual's role in determining meaning.

Clusters of meaning can be associated with the definition of home (Hayward 1975). A common concept of home is a physically defined structure that people sometimes inhabit. Home is also seen as a replaceable commodity that is bought, sold, and occupied. Home is associated with territory where occupants have a sense of intimacy and control; it is a series of territorial boundaries starting with the most private spaces, such as those where one sleeps, and moving outward to include those areas where there is a feeling of familiarity and belonging, such as the neighbourhood and home town. Home can also be conceptualized as a locus in space, a central point of reference in a person's daily life, where a person starts and to which he or she returns. Home can symbolize the essence of self and self-identity, expressions of myth, and idealized memories. It is a pivotal point around which human activities revolve and significant experiences occur. As an archetypal image, it embodies rootedness. Home can also mean a social and cultural unit that the family or community depends on for physical and psychological support, as well as a setting for socialization and acculturation.

Rapoport has argued that "home environments are part of larger, culturally variable systems of settings (the house-settlement system) and are themselves best understood as that system of settings within which a particular set of activities takes place" (1985, 264). In this framework, there are no preexisting definitions for home environments; rather, the system of settings where a particular set of activities takes place must be described in order to arrive at the meaning of a given space. For example, in some cultures the home environment may include the neighbourhood, which broadens the meaning of home to include a social system larger than the household.

The home has a variety of meanings in the modern era: control, security, refuge, orientation, comfort, entertainment, solitude, memories, accomplishment, family, children, space, personalization and expression, responsibility, investment, and seclusion (Edelstein 1986). As society has become increasingly complex, the physical manifestation of our belonging "somewhere" becomes more important. The search for home is a search for connectedness. The lack of rootedness described by phenomenologists "comes not only from the absence of a place to dwell, but also from having the dwelling experiences that constitute home cut from beneath one's feet by rapid advances in industrialization and technology" (Dovey 1985a, 48).

In North America the notion of home revolves around the "American Dream," in which the home provides both security and a focus for the identity of the nuclear family. To fulfil this dream often means acquiring commodities that make the household more efficient and self-contained. This has increased the tendency toward a more inner-directed home that offers protection from crime and the unknown elements of society. Only the very poor are forced to have dealings with their neighbours, because of substandard housing that necessitates interaction in public areas. For other groups there has been a convergence in space standards; middle-class and working-class homes differ only in tastes in decoration. The norm has become the single-family home with an open floor plan that imposes a certain pattern of living on the household.

The home in relation to its community setting can be interpreted as an arena where the varying socioeconomic, cultural, and personal aspirations of its inhabitants are manifested. These aspirations can come into conflict when, as with home-based work, new activities require the reorganization of the home environment. As the home changes to accommodate new patterns of working and living, maintaining a balance between these varying aspirations becomes increasingly difficult.

The Home As Buffer/The Home As Trap

Feminist theoreticians in planning and design (Markusen 1980; Hayden 1981) have analyzed the home environment in terms of the consumption and production activities of a housing unit. In these terms, the single-family home is the perfect vehicle of the patriarchal system under capitalism to increase demand by creating a need for duplication of goods and services. With the separation of home and work spheres, a spatial structure of isolated units connected to nodes of consumption was developed. Hayden explains:

> The private suburban house was the stage set for the effective sexual division of labour. It was the commodity par excellence, a spur for male paid labour and a container for female unpaid labour. It made gender appear a more important self-definition than class, and consumption more involving than production ... While the family occupied its private physical space, the mass media and social science experts invaded its psychological space more effectively than ever before. With the increase in spatial privacy came pressure for conformity in consumption (Hayden 1981, 172-3).

The electronic home is an extension of that trend. By transforming the home into a self-selecting telecommunications centre, the home is becoming both more insular and less private. As more information enters the home, occupants no longer have to leave the home to conduct many aspects of life, but at the same time it is difficult to screen out the enormous amount of information that is coming into the home. While teleworkers can turn off their computers and other technologies, they are still always present. For them, the home is losing its nature as a refuge as work-related stresses become associated with it. Effectively, teleworkers are wired into the job all of the time.

Though there is some evidence that homework, in general, affects the social and spatial patterns of the home, there is no clear evidence of how telework per se is changing the role of the home. Public life is intruding into the home more than ever as information can be received and social connections can be made electronically, without having to go outside the home. Information and telecommunications technologies allow information to be received and disseminated, and social connections to be made, independent of any specific physical locale. Whether this is changing the meaning of the locale for its users still needs to be considered. The impact of information technologies on the meaning of places has yet to be fully articulated. This will be further discussed in Chapter 7.

The home becomes utilitarian when the work done in it invades every aspect of home life. It is an isolating experience when working at home severs social ties. In contrast, the use of information technologies can enrich family life and broaden the concept of home by affording opportunities for the integration of home, work, and community in a localized setting. The critical factor is the kind of control an individual has over the activities generated by information technologies.

Technologies themselves do not transform home environments but rather support the decision-making processes that affect the activity patterns of the home. Technology supports changes in individual choices, family structure, and the relationships between the home and the school, the office, and other societal institutions. As Figure 6.5 illustrates, there are numerous modes of electronic communication that affect social relationships and activity patterns. For example, a data processor working at home and electronically monitored by her employer has limited control over her activities. To her, the home is confining and is diminished as a personal expression. In contrast, a self-employed teleworker who is hooked up to the Internet finds the technology liberating in allowing him to gather information and communicate with colleagues and others. He is free to engage in a variety of activities during the course of the day, including using the Internet for recreational purposes. Though he usually works long hours, he feels in control of his work and his environment, a situation that enhances his appreciation of his home.

While there are distinct class and gender differences among at-home workers, homework, in general, has created new relationships between the home and society. Horwitz contends that previously "work accomplished in the home, including household care, [tended] to be 'pre-industrial' – measured by the task and not by the clock, interruptible to the rhythms of human needs, and rarely rationalized or regulated by economies of scale" (1986, 194). The conflicts that have been described in previous chapters between home and work are reflections of the dual modes that are operative in the home.

In resolving the conflicts between the preindustrial and industrial modes of operation, teleworkers are redefining themselves as workers. While they are constrained by the pressures of corporate society and often are exploited because of their vulnerability, in many ways they have gone beyond the requirements of a worker. They have a new social identity that totally blends their personal and work life. This new identity may, in many ways, be coming full circle to the old "putting out" system of the medieval cottage workers described in Chapter 1. Under these circumstances people worked long hours, and domestic activities and space were mixed with work.

Doing paid work at home means the home is no longer a buffer from the stresses of the outside world. Instantaneous telecommunications also make the home less a refuge and more utilitarian. The public/private dialectic of the home becomes blurred as the home becomes integrally and visibly linked to the socioeconomic forces in society. While the home has always been linked to its context, the illusion of self-sufficiency has been perpetuated. The experience of teleworkers has made the contradictions of that illusion evident.

Many of the teleworkers interviewed, however, appreciate the control that this new role allows them. Most teleworkers have adapted their work to

Figure 6.5

Communication typology of the electronic home

Organization	 Home-generated decentralized	 One-way centralized	 Two-way decentralized	 Two-way centralized	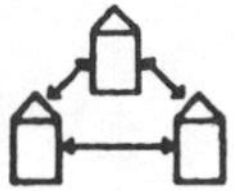 Network decentralized	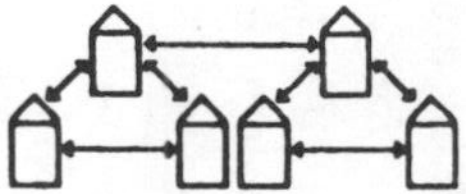 Network centralized
Technology	Audio systems Microprocessors in appliances VCRs Videodisk players Video games Hand calculators Stand alone PCs	Television Radio Cable TV converters Satellite TV receivers	Telephone PC with modem Facsimile	Local Area Network (LAN)	Local Area Network (LAN)	Local Area Network (LAN)
Services				Teleshopping Electronic banking Public videotext	Interactive cable Electronic mail Telecommuting Public bulletin board	Teleconferencing
Social relationship	Individual	Individual-global	Individual-individual	Individual-corporation	Individual-network	Individual-corporation
Social activity	Leisure/work	Leisure/consumption	Leisure/work	Consumption/work	Leisure/consumption/work	Work

new rhythms based on personal priorities. Many feel that they are their own bosses. At the same time, individuals working at home no longer have the support of fellow workers and are often divorced from an extended network of family and friends. In addition, teleworkers have no refuge when the pressures on them become too great.

With the loss of the home as a buffer, the stresses and demands of everyday life invade the home. The illusion that these demands can be controlled once they are in the sanctity of domestic spaces has not been borne out by the evidence. While the teleworkers interviewed varied in the degree that they allowed their work to usurp their home activities, some are making choices not on the basis of what kind of home environment they want to work in, but rather, what kind of work environment they want to live in.

7
Convergence: Telework
As Everywhere, Every Time

Heaven is the anywhere, anytime office. Hell is the everywhere, every time office (Paul Saffo quoted in Patton 1993).

For many years I worked for companies in offices outside my home. Then I turned my home into a workplace when I became self-employed. I've got a Dot Com. I am a workplace (Barbineau 2000).

Telework has a critical attribute that differentiates it from other types of work organization: the use of portable information and communication technologies allows it to be "delocalized," that is, work is free from the confines of a particular physical work environment. The aspatiality of the new technologies has created a distinct phenomenon: when a corporate identity requires only an e-mail address and a Web site there is no need to be physically based in any particular location. In the course of a day, work can be conducted in a home, in a car, at a client's office, in a restaurant or café, walking on a street, literally anywhere. An evocative photograph captures the aspatial quality of new technologies. Taken during the annual COMDEX convention in Las Vegas, a major computer convention, the photograph caught a number of people intently talking on their cellular phones on a wide street, oblivious to their surroundings and to each other. By their eyes you can tell that their physical setting holds little significance for them; the space that they occupy electronically is paramount.

In the ten years since this research began, the character and extent of computerization has changed significantly. In the first study only a handful of those interviewed actively used the Internet. By the mid-1990s that number, as evidenced by the percentage of those connected in the Canada-wide survey, had increased substantially. By the start of the twenty-first century the nature of computer work has been transformed to such an extent that the people I interviewed for the Vancouver study were conducting their work entirely on-line. The Internet was not just a tool to assist them in doing their job, but the source of their income.

Computers have clearly changed existing corporate cultures. Zuboff (1988) researched changes in the workplace with the advance of computerization, and identified several dilemmas that occur as offices are reorganized to accommodate office automation. The key question is whether computer usage will facilitate decentralization or centralization of decision making and information. While enthusiastic advocates of computerization maintain that computers allow for the democratic restructuring of corporations, critics are concerned that telework, among other trends, leads to fragmentation, not more egalitarian organizations. Wexler's (1999) research on business ethics articulates this dichotomy in the new world of work from the "postmodern optimists" and the "postmodern pessimists" perspective. The optimists see hierarchies being replaced by teams and rigid management modes giving way to strategic participation of workers to allow flexible responses to changing demands and values of the business world. The pessimists see these changes as a kind of ethical chaos in which fairness becomes relative and only the most vocal are heard.

Corporate culture pundits laud the new economy, dependent on new technologies, as the vehicle for the growth of political, economic, and technological openness and interdependence (Tapscott and Caston 1993). They encourage workers to "think outside the box" as a way of increasing creative problem-solving and redefining work satisfaction. The contradictions behind the publicized attempts of the business world to create more humane work environments become apparent, however, in light of the increased merging of gigantic corporate entities. In this framework, the new economy is viewed as the instrument for extending monopoly control and making it invisible through "virtual" organizations.

The previous chapters have sketched the new spatial and temporal relationships that teleworkers have to their workplaces and homes. In this chapter the consequences of these relationships for the meaning and role of workplaces and communities is further explored. At-home workers and mobile workers are evolving attachments to their physical locales that reflect their autonomy in where they work and live. The nature of this autonomy, however, needs to be deconstructed. Autonomy implies self-sufficiency and independence. While the new worker is self-sufficient and independent to an extent, he or she is also inextricably entwined with and reliant on the new technologies, and correspondingly vulnerable to them.

The convergence of the computer, television, and telephone allows these communication tools to offer increasingly broad coverage. However, the penetration of media into every facet of our lives, especially with our increasing reliance on the Internet for information, entertainment, and commercial transactions, means that we are vulnerable to public scrutiny in the most private aspects of our lives, including our work patterns, buying habits, and choice of entertainment. Convergence, a seemingly benign term, entails a whole host of technological dependencies that affect our daily lives. In the context of telework it also implies the increasing lack of differentiation between our work lives and our domestic lives. The social meanings of home and work environments are beginning to merge as they become informationally similar. Who controls the information, the technology, and their use, however, are the significant issues.

Virtual Workplaces/Virtual Companies

In 1989, Rick, thirty, a technical writer, believed he was on the cutting edge of the use of telecommunications and information technologies. He provided technical support at Apple Computer, in California's Silicon Valley. A committed telecommuter, he rarely went into his company's offices, but worked at home in his living room using an easy chair and a low table for his computer, modem, and telephone. His workstation consisted of a computer, a hard disk drive, a CD-ROM player, a high-speed modem, two telephone lines, a printer, and a program that allowed access to electronic mail. He was single and lived alone in a one-bedroom apartment in a barrack-like housing complex in the garden of a larger house in Palo Alto. He had two dogs.

He called his home workspace a "virtual workspace," a borrowing from "virtual reality," which relies on interactive computer graphics to create the illusion of navigating through artificial locations that (in the hypothetical ideal) seem as real as those of the real world. Similarly, working at home, Rick could create the illusion of a workplace that is indistinguishable from a "real" office. Because he worked and communicated electronically, he worked in "electronic space." No one really knew where he was working, and no one really needed to know, because he didn't need face-to-face communication with his employers. The other eight people in his work unit also telecommuted. They had regular all-day group meetings once a month, rotating them in each other's homes. Rick felt real bonding took place in these meetings. They had breakfast and lunch together, and got to see each other in their own milieux.

Besides Rick's full-time employment at that time, he was a partner in a company that produced videos and was enrolled in a film production class at an art school. His schedule allowed him great flexibility. He got up early so he could make long-distance calls to clients, and worked until 11:00 a.m. From 11:00 to 6:00 p.m. he was involved in his other activities. After dinner he worked until midnight. He put in eight hours spread out over the entire day, at times that suited him best. He worked six to seven days a week, and made no distinction between weekdays, weekends, and holidays. Though he acknowledged that he was very lonely in his immediate neighbourhood because nobody was at home when he was, he got over his loneliness by using the telephone.

Rick characterized himself as a workaholic. A college dropout who had done a variety of odd jobs, he was "discovered" by a lawnmowing customer who managed a technical writing group at Apple Computer. Hired by this man as a technical writer, Rick proved adept at it and soon was made a manager. After three years, finding himself overworked, stressed, and overweight, and recognizing tendencies like his father's, who had died of a coronary arrest six months before retirement, Rick asked to be demoted to a technical writer again. Soon after his demotion he asked to experiment with telecommuting. Initially very uncomfortable with staying at home, after the first month he evaluated his output and discovered that it was five times higher than when he worked at the corporate headquarters. Without interruptions he was able to work for long periods of time. With the autonomy to set his own work schedule, he changed it to match his body's natural energy cycles, working mostly at night or in the early morning. With the time available to him, he rebuilt his social life, which had disappeared while he was in management, and began to pursue dormant interests such as acting.

Rick was firmly convinced that the capabilities of telecommunications and information technologies had freed him to live the kind of life he wanted. He felt as if he was "homesteading" on the cutting edge of transformations in the workplace. To Rick, the office was an inefficient and unstimulating environment for creative interaction. In the future he envisions a completely decentralized work organization with no need of a centralized office facility. Most people will telecommute, and corporate entities will be obsolete. Instead, people will form "virtual companies," fluid organizations that will be set up for specific projects and then disbanded after their completion. He, and many others, observe that companies are downsizing corporate headquarters as part of corporate restructuring.

Since 1989, flexible work organization has become increasingly the norm. Small businesses are relying on staff who are willing to offset security for flexibility. Judith, fifty, is the principal owner of such a virtual company. Her company does market research and product development for a variety of companies. She works from her home in West Vancouver, a suburb of Vancouver, and has a staff of eight who also work from their homes, including her secretary. They often communicate over the computer rather than having face-to-face meetings. There are a large number of other people that they also employ depending upon the work. As she describes it they are constantly forming and reforming work groups for specific projects, creating fluid and flexible organizations. All of her staff are women, and most are highly trained professionals, many with graduate degrees, who opted out of a structured organization because they needed their work to accommodate raising a family. They all work as independent contractors for Judith.

Derek, profiled in Chapter 5, also believes he is on the cutting edge of new work organization. As a technology support consultant based in Vancouver, he is currently working on a project in Los Angeles with a "virtual team," a group of people from Los Angeles, Seattle, and Vancouver. The team communicates through e-mail, telephone conference calls, fax, and voice mail, and every few months they go to Los Angeles for two or three days at a time. This fluid work organization suits the needs of these workers. They don't have a goal of permanent team-building, because they recognize that they will disband at the end of the project and go on to other work. However, this may not be the best situation for companies trying to build teams. People working on short-term contracts are very expensive and in a volatile industry it is easy for them to find other work.

In an exposé of the new work frontier a 1993 *New York Times* article described an advertising agency with a totally decentralized work environment (Patton 1993). The article chronicled a day in the life of Denise, a female executive in charge of several important accounts who was part of a pilot project to adjust patterns of work in her organization. The goal was to create a "virtual office": an office without walls. She had free rein to work almost anywhere – at home, in the company's conference room, or on the phone in the front seat of her car. In the back of her car was a cardboard box full of family photos and other memorabilia that had once decorated her office at the advertising agency. She had no natural place to put them.

Her company, the advertising agency Chiat/Day, based in Venice, California, hopes that working this way will "empower" its employees, giving them more responsibility for where, when, and how they do their work. More importantly, corporations such as Denise's, which have traditionally invested in real estate for large corporate headquarters, see the opportunity to reduce overhead costs. Many large corporations are now basing their long-range planning on a dramatic reduction in office space. Recognizing that with new technology the workplace can be anywhere, they are organizing their corporations around a mobile workforce.

Denise's company's headquarters underwent a renovation in 1989 to make it "virtual." Organized around the concept of a campus, rather than corporate offices, there are project rooms resembling seminar rooms, a media centre, and staff work and socializing areas, similar to a student union. Rather than individual office cubicles, each employee was given a locker to store personal items. Upon entering the building a worker stopped at the "company store" for equipment such as a writing pad and laptop computer, then moved on to the "student union" where a concierge registered her for a work space and where she programmed her telephone. Alternatively, she might do research in the media centre, lined with video equipment, computers, and CD-ROMs, and perhaps end her day by organizing a meeting in one of the project rooms.

To survive in an uncertain economic climate, corporations have had to become adaptable and flexible. Through downsizing, corporations are becoming little more than shell operations, employing a handful of people to do one strategic thing and contracting out other work. Recognizing that changing work conditions necessitate new models of office organization, corporations are looking at building types that reflect mobility and change, such as hotels, campuses, and airport lounges. In these new corporate cultures, workers are "hotelling": setting up offices for temporary use, like hotel rooms. The staff are divided into two distinct types of workers: "teamers," the employees who spend most of their time in the office, and "mobiles," those who spend most of their time away from it. Clearly there is a division between these two groups; the former could be seen as the permanent staff, while the latter are transient, floating personnel (Patton 1993).

While advocates of this new office organization extol its virtues, the virtual office may not suit all employees. Many would miss the social fabric and human contact of the office environment. To address this, proponents are talking about finding a "virtual water cooler," a device, perhaps a videophone, to keep those working at home linked to the camaraderie and gossip of the office.

The virtual office is, however, not a specific place, but a process. It is "a bubble of information created by new technologies such as the cellular phone, the laptop computer, the modem, and the new generation of smart beepers" (Patton 1993). For Rick, Judith, Derek, and many others, the role of the workplace as a physical place where personal attachments are made has been replaced by the sustenance they derive from their "electronic spaces." Now they must choose their living situation based on their work requirements. However, they are not just living in their offices. Their sense of what entails their home and work spaces has become totally blurred and enmeshed in a newly defined space that is not physical.

Many teleworkers cite the hostile office environment as a significant reason for opting to work at home. Working at home is a retreat from negative social and physical factors in the office. Teleworkers hated the gossip and "back-biting." Many felt they were in dead-end jobs that were only exacerbated by close proximity to their bosses and coworkers. They couldn't work efficiently in an office setting because there were too many distractions. Rick, the technical writer and committed teleworker, elaborates: "The office is an inefficient and inappropriate place to do any form of human interaction. The office is inherently hierarchical and based on power relationships, but for many people it is their only 'family.' The office creates psychological ruts where there is a group think mentality which inhibits individual creativity." He goes on to describe the physical office environment: "Office buildings are designed deliberately as 'human isolation tanks' so people can be separate from the world. The soft lighting, quiet, carpets – everything in the office is there to isolate and make you concentrate on work. But you can't work in them because they are terribly unhealthy. I believe that bad ideas come out of these environments."

Most of the office workers interviewed recognize the problems in their work environments. The most commonly cited problem is the open office plan with partitions, rather than walls, to separate workspaces. The majority of part-time teleworkers and office workers interviewed in the California study worked in an open plan office in a high-rise commercial building. In these environments there is no acoustic privacy, and often there are too many people in a space. Everyone can overhear conversations, which can be distracting and disruptive to the rest of the office. Most office workers disliked the fluorescent lighting, which creates glare and gives them headaches. A few had severe allergic reactions to the synthetic carpets and the recycled air of the high-rise environment.

Those whose employers require them to work at home part time feel even more disconnected from their office setting. They regard their corporate office as a much less productive place to work. They have no designated office workspace and share computers and other equipment with other workers. They have no privacy and are constantly bombarded with external stimuli that disrupt their concentration. They feel like intruders in this environment and recognize that their office-bound coworkers resent the flexibility of their work situation and the inconvenience of having to share their resources when the part-time workers are in the office.

Though there are many problems with the office environment, the office workers interviewed in the California study prefer to work there than at home. In many ways the office is their second home. They socialize in it, and they feel connected to the world through their workplace. They like the distractions, even though they impede their work. Most office workers have personalized their office space: family portraits and pictures reflecting their tastes adorn walls or partitions. A few offices even have homelike furniture, such as sofas, for relaxing after a long day. For these workers the office has a strong personal attachment. Their social identity is reflected in its environs.

In a *Vancouver Sun* article (Emmerson 1999) on what makes IT jobs attractive, the author quotes an employee of Electronic Arts, an international new media and gaming company located in Burnaby, who describes the perks provided by his employer on site: "EA provides cheap gourmet lunches, free overtime dinners, a gym, steam room, martial arts classes, big screen televisions, pool tables, a lounge area with a fireplace, giant decks with fantastic views and all the mind-numbing video games you can play." These amenities seem surprisingly similar to what you would find in a home. For these workers, who work long hours, home is often no more than a pit stop to change clothes, sleep, and occasionally eat a meal. A realtor describing a development catering to high-tech firms where people can live and work comments, "People spend so much time in their offices these days, they want everything right there" (Bula 2000).

The idea that work is gaining significance as a place where a whole range of needs for camaraderie and identity are met is now being documented in academic research as well as the popular press. In her book on the strategies of working parents, Hochschild (1997) describes work as the sanctuary to which a growing number of workers are fleeing in an effort to escape the time demands imposed on them at home by children and spouses. Their identity as competent, accomplished members of society is reinforced at work, not at home.

While those who choose to work in an office setting appreciate its benefits, almost all of the teleworkers, both full-time and part-time, regard the office they left (or the one they go to part time) as a less efficient place to do work. Now that new technologies and more flexible corporate management have made work activities independent of a particular physical setting, these workers have opted for an alternative to their stultifying workplaces. For them, their home, even with its problems, is a more conducive environment for their work life.

Community Sphere

What effect does this aspatial relationship that teleworkers have to their workplace have on community life? The neighbourhood has been perceived by planners and urban designers as a physical locale where residents are in face-to-face contact with each other because of their proximity. My research has found, however, that urban North Americans do not live as envisioned by planners. They have limited contacts with neighbours, and their important social contacts are with friends, workmates, and family who may be thousands of miles away. Home-based workers are no more prone to interact closely with their neighbours, nor to perceive their neighbourhood as their community, than office workers. Nevertheless, homeworkers have a greater perception of the isolating nature of their immediate environment.

This total isolation was not felt by all of the homeworkers. Over one-third of the California sample and close to half in the Canada-wide survey live in an urban neighbourhood, and over one-quarter in both samples live in a mature suburban neighbourhood (i.e., a neighbourhood that was built originally as a suburb but has since been infilled with services and denser development due to the growth around it). For them, services are relatively close, though few know their neighbours. Nevertheless, for the one-third who live in a suburban neighbourhood, and the one-fifth of the respondents who live in an exurban setting, the lack of social contact coupled with the lack of services makes their neighbourhood environment a lonely place.

The more residential an area, the more potential tensions and problems may be encountered in integrating home-based work there. Over four-fifths (83 percent) of the respondents in the Canadian study describe the character of their street as residential. Only 8 percent identify their street as mixed-use, while another 1 percent describe their street as commercial. Eight percent describe their street as rural. Almost all the respondents have shops and services within a ten-minute walk or a five-minute drive from their homes except for copy centres, recreation centres/gyms, and child care centres, which are at a much longer travel distance.

For those interviewed in this research, the perception of what constitutes a neighbourhood varies greatly according to the type of neighbourhood a person inhabits. Urban dwellers perceive their neighbourhood's boundaries in terms of walking distance to services (though they may not actually walk to these services). This usually translates to a four-block radius. Rather than distance to availability of services, suburban dwellers perceive their neighbourhood's boundaries in terms of identifiable borders, such as those of the housing development in which they reside. A few suburban dwellers perceive their neighbourhood as an area in which they are on nodding acquaintance with their neighbours. This usually translates to a one-block radius. Though they may not know most of their immediate neighbours, their children usually play with other neighbourhood children in this area. There does not seem to be any difference in perception of neighbourhood boundaries between homeworkers and office workers from the California study.

While most of the respondents live in neighbourhoods where they are in close proximity to their neighbours, when asked about neighbouring patterns, few respondents claim to know many of their immediate neighbours well. Jenny, the medical transcriptionist in Hercules, California, is surprised at how lonely she feels working at home. Her suburban community is empty during the day, since most of her neighbours are in dual-career households and work elsewhere. She rarely sees these neighbours because they usually drive their cars into their garages and enter their houses from there. To go where she will see people requires driving. Tim, the teleworking creative director in Vancouver, finds that although he lives in a dense community where he knows a lot of people, "I have a stronger awareness of my community but I don't socialize more in it. When I am home working, I am in a shell."

There are no appreciable differences between homeworkers and office workers in their use of their neighbourhoods and their services, since neither group has many social contacts in their neighbourhoods, nor do they do many activities in their immediate communities. This finding corroborates a pioneer study on home telecommuters by Olson (1983) that reported that the relationship between homeworkers and their communities had not been altered by the time spent working at home. They did not become more involved in community activities, although they did spend more time in leisure activities.

Both homeworkers and office workers in the California study, on the whole, did not use the services in their neighbourhoods, either because there were no services available, or because they preferred using them elsewhere. Male home-based workers and office workers used services the least, while female homeworkers used services the most. Interestingly, both male and female part-time homeworkers used services similarly, generally more than male home-based workers who were at home full time. The services that were used the most were the shops, though many married male homeworkers never used these either because, as one acknowledged, "My wife does the shopping."

Most neighbourhoods lacked places for informal socializing, such as cafés and/or restaurants, but when they were available, they were used. Recreational facilities were rarely used in the neighbourhood. Both homeworkers and office workers preferred to go to recreational activities of their own choosing, not just those that were locally available. Homeworkers, especially, felt the need to get out of the immediate environment. When asked whether they used neighbourhood services more when they were working at home, the majority of homeworkers said they used their neighbourhood services the same amount. One teleworker especially emphasized: "When I am at home to work, I work. I don't have the time to go out and do other things."

One at-home worker has dealt with the issue of isolation by deliberately choosing to live in an environment that provides him with a community. Dick, forty and single, believes he has found the right combination of privacy and community in his immediate environment. A self-employed architect, he works in a small two-and-a-half-storey house that is part of a small cluster of homes in an inner-city neighbourhood in Oakland, California. Although the houses are individually owned, the courtyard and garden are shared. Dick's workspace is on the main level of his house and has a separate entrance to the courtyard (Figure 7.1). Upstairs, his living space consists of a combined living/dining/kitchen area and a small bedroom loft.

Because Dick lives in a community where he knows his neighbours intimately, he rarely lacks for company. He is able to have daily contacts with his neighbours. He doesn't distinguish his home life from his work life, and he believes that he has found a positive balance between the two. He acknowledges that he is not out "in the world" that much, but he isn't living alone either, because he lives in a small community where other people also work at home. He describes his life this way:

> I always wanted to work at home. I always wanted the independence and freedom of being my own boss. I like that my work is right downstairs from my home and near the courtyard – near people that I care about ... In the compound there is a lot of informal interactions, not formal socializing which is great. If I want to talk with someone, it is OK, and if I don't want to talk, that is OK too. This community is very important to me in meeting my needs as a homeworker but I do miss the camaraderie of the studio.

Figure 7.1

Site plan of homeworker's housing compound

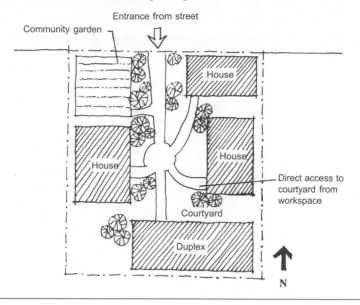

Important to Dick is how people interact in the common areas between their houses. He doesn't have much personal contact with people outside of his compound, but he does know people on the street.

The majority of homeworkers don't live in such an environment. There is very little common space in most residential neighbourhoods. Few home-based workers have daily contact with their neighbours. They rarely notice what is going on in their neighbourhoods, the presence of strangers on their street, or children playing. Many have workspaces that don't front the street, and they have to keep windows shuttered to prevent computer glare. Moreover, most are too involved in their work to notice. This finding corroborates Ahrentzen's 1987 study of homeworkers. Her study found that computer technology decreases a person's likelihood of looking out the window, and that generally, her sample did not notice what was going on in the street. Those in my research who do notice neighbourhood activities find that not much is going on, and that they have had to get used to the quiet. They are somewhat apprehensive of strangers in their neighbourhoods, if they observe them at all.

In the California study, homeworkers do not socialize with their neighbours more than office workers. Their neighbours are usually at work elsewhere. Only a few have casual relationships with their neighbours. In addition, homeworkers do not spend more leisure time in their neighbour-hoods than they do when they are not working at home. The leisure time they do have, they tend to spend away from their immediate locales. This, again, corroborates Ahrentzen's study. The majority of homeworkers in her study did not recognize more of their neighbours since they began working at home, nor did they feel a strong attachment to their neighbourhoods.

Homeworkers generally perceive their neighbourhoods as very lonely places with few social opportunities. When the respondents in the Califor-nia study were asked what improvements would make their neighbourhoods better places for them to work at home, only a few had suggestions. Some would like a café where they could go for lunch and talk to people. Most, however, had no suggestions because they couldn't make the connection between the isolation of working at home and neighbourhood improve-ments that could alleviate some of that isolation. This connection was hard to make because they did not feel they had much control over improve-ments in their neighbourhoods. Development of the neighbourhood was perceived negatively in terms of increased density and traffic. While many were not satisfied with present conditions, at the same time they did not want their neighbourhoods to change. Canadian respondents, on the other hand, suggested various improvements to their neighbourhoods, including better transit, improved local shops and services, and improved telephone services.

Use of the Community
Like their California counterparts, home-based workers in the Canadian sample have not changed the use of their neighbourhoods since working at home, nor do appreciable numbers interact more with their neighbours. They do generally feel, though, that working at home enhances the security of their home and neighbourhood. As well, for some home-based workers the community has become a great source of friendship and support. Over one-quarter of the home-based workers would be interested in working from a neighbourhood telework centre or satellite office. They see the advan-tages of being close to home without the distractions of home, and would appreciate the services and camaraderie that such centres would provide.

Generally, the Canadian respondents use the services in their communi-ties about the same amount as when they are not working at home or be-fore they worked at home. On average, respondents use the post office, the copy centre, and banks more when they work at home. However, over two-fifths (41 percent) of the home-based workers use cafés, restaurants, recrea-tion centres, and gyms less when they work at home, and nearly one-third (31 percent) use parks less when they work at home. This reflects a general lack of time as well as the priorities of home-based workers regarding social and recreational activities. As discussed in Chapter 4, over two-thirds of the respondents (69 percent) spend their leisure time close to home when they work at home. Thus while home-based workers in this sample are more likely to do leisure activities close to their home when they do them, they do not appear to engage in leisure activities more now that they work at home.

When they work at home, 43 percent of respondents notice what is happening on their street more than when they are not working at home or before they worked at home, while 42 percent notice street activities about the same amount and 9 percent notice these occurrences less. Close to two-thirds (64 percent) interact with their neighbours the same since working at home, only one-quarter (25 percent) interact with them more, and 7 percent interact with them less. This data appears to contradict the data presented in Chapter 4 that half of the respondents know their neighbours better when they work at home. However, home-based workers may know their neighbours casually as they see their comings and goings but choose to not interact with them on a regular basis. This level of interaction is common in North American residential settings with high mobility. As well, home-based workers may lack the time and inclination to engage in interactions that will detract from their work.

A major planning concern about home-based work is that it will adversely affect the residential character of neighbourhoods by increasing traffic and introducing uses that are not compatible with residential living. These concerns are not reflected in the findings. Two-thirds of the sample feel that the work-related activities in their home have no effect on the residential atmosphere of their neighbourhood, while over one-fifth (21 percent) of the respondents feel that it has a positive or very positive effect, and only 2 percent feel it has a negative effect. One respondent who feels it has a positive effect states, "I see my neighbours more often, I go to the post office every day and thus meet many people in the community." Other comments include, "Neighbours know I'm more available if needed, but they don't really know whether I'm working or 'at home' but I am physically present." Another believes that home-based work is making a positive contribution because, "I am in and around the neighbourhood more, thereby enhancing its security." An artisan believes, "The community is delighted to have me and other artists a part of it," and another comment concurs, "Many neighbours work at home. We see each other often, [creating] a great appreciation of what constitutes work."

Others believe that their home-based activities have no effect on their communities. One home-based worker states, "When working at home I'm very focused, thus I spend most of my time working, not walking in the neighbourhood" and another writes, "Sure my neighbours are not even aware that I telework." Still another states, "My work does not affect the residents around me. No one is aware that I work at home. The houses are set far apart. Everyone keeps to themselves and I prefer it this way." Another home-based worker concurs, "Since I have no extra traffic as a result of being at home, I don't think it has affected our neighbourhood."

Relocation to Neighbourhood Telework Centre or Satellite Office

A significant number of respondents expressed interest in working either from a neighbourhood telework centre, which is a place with workspaces shared by a number of unrelated businesses and located conveniently in a neighbourhood, or a satellite office, which is a company's secondary office located close to employees' homes. This supports the argument of Johnson (1999) that a neighbourhood telecommuting centre could offer environmental benefits without the social costs of home-based work. Over one-quarter (27 percent) of the respondents in the Canadian survey would be interested in working from a neighbourhood telework centre or satellite office site, while 56 percent wouldn't be interested, 10 percent don't know, and the rest don't find it applicable to their work situation. Teleworkers and independent contractors were the most interested in working in a neighbourhood telework centre or satellite office, while self-employed consultants and home-based business operators were less enthusiastic. Half of the teleworkers in the California study would be interested in working in a satellite office near their home. Two-thirds of the female independent contractors and self-employed entrepreneurs, but only one-third of their male counterparts, would be interested in working in this arrangement.

Those who are self-employed in the Canadian survey would be interested under certain conditions. As one woman commented, "It would depend where the centre would be located. I don't want to start travelling a long distance again." Another woman entrepreneur concurred, "Yes, if it were some place that could take messages. I would have use of fax, photocopier, computer, etc. There would be more interactions with people with all the office facilities, but commuting would be diminished. It offers the best of both worlds: flexibility plus less distractions from home." Male home-based entrepreneurs had similar comments. As one noted, "It would be an advantage in marketing business services and provide stability that would not cost very much," and another found, "The idea of a common workshop is intriguing." One man was quite adamant: "I want more human contact. I could share cost of equipment, copier and someone to answer the telephone. The occasional use of a board room would be nice." A telework centre could reduce costs by sharing resources without giving up flexibility, as well as providing an opportunity to be with others. As well, availability of services and visibility would be increased.

Those who would not want to work in a neighbourhood telework centre find that they work better at home and don't see the cost savings. They are concerned about losing the flexibility they have at home to set their own hours, and not being close to their families. For them, the benefits of working at home outweigh the disadvantages.

Travel Patterns

Telework, if it becomes widespread, is seen as a potentially significant factor in reducing future automobile use by shortening or eliminating the trip to work. The findings in the Canadian survey do demonstrate a decrease in trips. However, travel patterns were not the main focus of this research, and data collection on this issue was not very sophisticated compared with other transportation-focused evaluations of telecommuting programs, such as the Puget Sound Telecommuting Demonstration (Ulberg et al. 1993) and the State of California Department of Transportation research on the transportation impacts of telecommuting (Mokhtarian 1991b).

For this sample, working at home does not change the preferred mode of transportation, the automobile. Almost all of the respondents (95 percent) use an automobile regularly. However, over one-third (35 percent) use their automobiles less when they work at home, and 30 percent use it the same amount. Interestingly, over one-third (35 percent) use their automobiles more when they work at home. While teleworkers use their automobiles less when they work at home, self-employed homeworkers whose work requires them to visit clients and suppliers use their automobiles more (Figures 7.2 and 7.3).

Figure 7.2

Transportation use of self-employed homeworkers since working at home

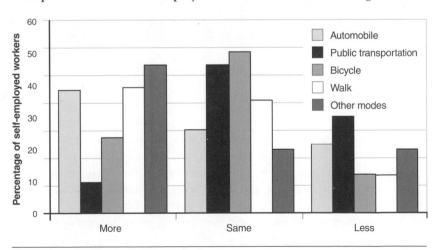

Source: Reformatted from Gurstein (1995).

Figure 7.3

Transportation use of teleworkers since working at home

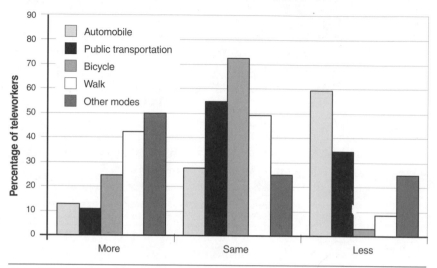

Source: Reformatted from Gurstein (1995).

The number of respondents who use public transit is significant, and surprising since they also use their automobiles regularly. Of the 202 respondents (45 percent of the total) who use public transportation, over half (55 percent) use it the same amount since working at home, 34 percent use it less and only 11 percent use it more. It appears that the majority use public transit only occasionally.

Three-quarters (76 percent) walk regularly. Of those, 45 percent walk more when they work at home, another 44 percent walk the same amount, and 11 percent walk less. Of the 226 who use a bicycle (half of the total) close to two-thirds (63 percent) cycle the same amount, 27 percent cycle more, and 11 percent cycle less. Bicycle use is still predominately recreational and not an alternative to the automobile to an appreciable extent. Some respondents reported that they used taxis more since working at home, and they also often use couriers to transport documents and other materials.

Data from the Canadian survey demonstrate that travel patterns of teleworkers differ from those of self-employed workers. The majority of teleworkers use their cars less and travel much shorter distances for all activities when working at home. In terms of the median distance travelled per day for work, household chores, and leisure, teleworkers travel significantly less per day when working at home. For self-employed workers the difference is less striking, since they travel a great deal in the course of their working day (Figure 7.4). It appears that reductions in transportation use

Figure 7.4

Median distance travelled per day by teleworkers and self-employed homeworkers

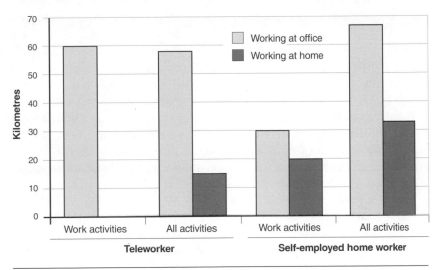

Source: Reformatted from Gurstein (1995).

realized through telework programs would be exceeded by the increase in use from the larger self-employed home-based segment of the population. Mokhtarian (1997) found that while telework may replace the traditional journey to work, other factors, primarily non-work trips, may induce replacement travel.

Because the self-employed often work away from home as well as at home, it is unclear whether, on the whole, they travel less when they work at home. More than two-thirds of self-employed workers are now using their car as often or more frequently since they began working at home. To accurately assess their net use of transportation, it would be necessary to obtain more detailed information on their travel patterns as they make deliveries and visit clients. Salomon (1985) noted that the interactions between transport and telecommunications vary according to the context and that substitution does not occur just because the technology is available.

Municipal Situation

While it is difficult to assess the extent of home-based economic activity, home occupations are an issue for municipalities. A study of greater Vancouver municipalities (Gurstein 1993), in which municipal economic development officers were asked to assess their current home occupations ordinances and the impact that home-based work is having on their communities, revealed a growing recognition among government officials of the importance of this economic activity. With the growing reliance of work on information and communication technologies, working in isolation at home has become a viable means of carrying on a business, reducing the high overhead costs of renting space and the need for transportation to and from the work site. Principal municipal concerns were: adherence to municipal bylaws, zoning ordinances, and regulations regarding space and employees; disruption and change in character of residential neighbourhoods; parking and increased traffic; unfair competition with commercial activity; and safety and security. Municipalities are concerned with adherence to the municipal bylaws and regulations because many home-based workers do not obtain business licenses and therefore are not regulated. Though most municipalities in North America have zoning regulations, bylaws, and licensing mechanisms to regulate home occupations, few have effective mechanisms that reflect current types of occupation conducted and technologies in use.

While few municipalities have comprehensive policies on home-based work activities (often called "homecraft") many municipalities in North America have found that they have to address live/work use. A live/work unit is a combination workspace and dwelling unit, such as a live-in studio for artists or artisans. Live/work spaces are a dynamic, evolving response to changing social and economic conditions. Policy documents and studies of these changing conditions recommend flexible and adaptable land-use policies that regard underutilized land in cities (especially industrial land and buildings) as a resource to encourage economic revitalization.

Live/work units can play an important economic role as job creator, job incubator, and job catcher. The Fish, Kurtin, and Nasmith study (1994) of waterfront regeneration in Toronto identifies the advantage of live/work design as its flexibility in responding to market shifts. However, most live/work policies tend to remove that flexibility by limiting live/work units to either artist or home-office space. This is usually due to conflicts with the building code when mixing residential with commercial and industrial uses. Fish, Kurtin, and Nasmith recommend that regulations rely on performance standards that restrict undesirable external effects. An evolving typology for employment/industrial units is recommended that recognizes three zones within each unit: public, private, and crossover, which may be either public or private, depending on the activity. The size of the zones and degree of separation between them will be determined by the particular mix of four elements: income, household size, lifestyle, and type of employment.

Other studies concur with the need for flexible land-use policies and regulations. Loomis (1995a) describes a mixed-use manufacturing and residential district in Brooklyn, which predates the separation of uses through zoning, as a model for urban industrial revitalization. The model is suggested as a way of dealing with vacant industrial and manufacturing spaces. In another study, Loomis (1995b) describes *hotels industriels,* flexible, mixed-use, multi-tenant facilities that primarily house production activities and are located within Paris' working-class districts. In 1978, in response to the deindustrialization of Paris, the government launched a program promoting these *hotels industriels* to retain existing production enterprises and to encourage new ones. Currently, there are 40 of these projects operating successfully in various parts of Paris. These districts mix ateliers accommodating such activities as printing, woodworking, video production, software developers, and graphics studios with housing (including a large component of social housing), daycare centres, shops, offices, schools, and a park. In many cases the city has facilitated development by providing city-owned land and expediting the permit process and other municipal regulations. Hatch (1985) describes a similar initiative in Italy that promoted the development of "working quarters" or "artisan villages" as an approach to economic revitalization that emphasized high quality and small-scale production.

In Canada, Orser (1993) investigated "hybrid residential development," a structure that is intentionally designed to incorporate both residential and business spaces and activities. Orser found few operational hybrid developments in Canada, suggesting that hybrid design is in its infancy. Her report suggests the need for more flexibility within zoning categories and the building code, providing for mixed use or residential/commercial use.

Zoning and Regulatory Restrictions
Home-based workers in the Canadian study generally feel that the municipal regulations governing their occupations are too restrictive. These regulations do not acknowledge the nature of much home-based work, which is non-toxic and non-hazardous and rarely has a major negative impact on the functioning of neighbourhoods. It appears that many home-based workers want their municipalities to reconsider their policies on home occupations, including revamping zoning and regulations to more accurately reflect the nature of their work. Municipal regulations need to be developed that differentiate between various types of home-based work activity.

The respondents who are affected by municipal ordinances are self-employed entrepreneurs who run a business or consulting practice from home. Teleworkers especially have little adverse impact on their communities, and zoning regulations are basically a non-issue from their perspective. Home-based business operators, depending on their type of occupation, may require clients to visit or small-scale crafts production or manufacturing. Consequently, they are more likely to be in violation of municipal regulations in residential neighbourhoods. While these non-conforming occupations appear to be only a small segment of the entire home-based work population, their presence is causing the most concern among neighbours and regulatory bodies. One-third of the respondents in the Canadian study have dealt with regulatory restrictions as a result of home-based work activities. Of those encounters, over four-fifths (82 percent) dealt with business licences, one-quarter involved inspectors, 21 percent dealt with building permit regulations, 20 percent dealt with sign bylaws, 12 percent dealt with zoning variance regulations, and 12 percent dealt with parking regulations.

Dealing with municipal regulations has not often been a positive experience. One female home-based worker commented on what she believes to be harassment by the city, "Because what we are doing is legal. There seems to be a sense that we're trying to break the rules even though we haven't," and another concurs, "A lot of red tape. We are not legally allowed to have clients in the home. This is ridiculous." Most regard their home-based work as simple, unobtrusive activity that shouldn't require meeting municipal regulations. Often they simply avoid reporting their business activities to municipal authorities. As one home-based worker commented, "I have avoided the whole issue by not applying for any permission," and another agrees, "I have not contacted them at all about my business. They would only increase my overhead costs for no useful purpose at all." Others have taken a more proactive approach by being instrumental in getting their local council to legalize home-based businesses and helping to draft new bylaws regarding their operation.

Respondents suggested that changes to zoning would make it more convenient to work at home. Most cited suggestions were the recognition of the non-hazardous nature of most home-based businesses by municipal authorities and allowing home-based work activities in the zoning bylaws. The comments included, "Zoning based on performance standards – not prescriptive single use land-based zoning"; "Fewer restrictions on licence – cannot legally operate my business as there is no appropriate category"; "Signage requirements changed to allow small business sign"; "Zoning to allow 'knowledge work' at home"; and "Zoning to allow retail sales space and other non-agricultural business activity." One comment specifically addressed the regulatory process: "It would be nice if the process one has to go through for zoning licence, etc. was done in a quicker, more efficient manner. Serious financial losses can [be], and have been, incurred by these delays."

Most single-use residential neighbourhoods have excluded work opportunities through zoning ordinances. Some residents are concerned that telework and home-based employment can generate heavy traffic and noise and place demands on parking. Many home-based workers are hesitant to identify their home as a work site because they fear that their neighbours will object. Some wish to keep the income they derive at home hidden from the government, and so would rather their work be invisible. If work-related functions are to be regulated in these environments, the zoning ordinances would require a rational system for classifying persons or property so that no arbitrary discrimination results. One approach is zoning based on performance standards, which assesses and encourages compatible uses in a systematic and inclusive way. By providing for home-based employment in official community plans, municipalities can regulate on the basis of these performance standards, rather than by the prescriptive definitions found in zoning provisions.

Zoning ordinances could also differentiate between major or high-intensity and minor or low-intensity home occupations (Butler and Getzels 1985; Frank 1993). A minor home occupation is one in which no persons other than members of the household are engaged in the occupation, there is no visible exterior evidence of the conduct of the occupation, it does not create a need for off-street parking, it does not generate additional traffic, and there is no equipment other than that in household, domestic, or general office use. These occupations, which include accounting, word processing services, consulting, and graphic services, may not necessarily require a special use permit as there is minimal impact to the neighbourhood. A major home occupation, in which people other than resident members of the household are employed on the premises, there is a sign, and both dwelling and home occupation parking needs are accommodated off the street, may require a special use permit and notification to adjacent neighbours regarding the change in use because of the potential impact on the neighbourhood. This type of occupation includes repair shops, light manufacturing, and services such as beauty salons and doctors' offices.

Another way of categorizing home occupations is by type of activity (Orser and Foster 1992). These discernible categories of home occupations include labour-intensive (e.g., renovation services), knowledge-intensive (e.g., architecture), and technology-intensive (e.g., telecommuting). These categories relate to the community impacts associated with home occupations. For instance, labour-intensive occupations create odours, noise, and visibility problems, while knowledge-intensive occupations may be high traffic generators. Technology-intensive occupations tend to be favoured because they are generally "invisible" (i.e., they don't cause noise, odours, and traffic).

Another approach to categorization depicts home-based work in terms of a typology that differentiates among types of work and the degrees of balance between living and working. This typology and the uses that correspond to these categories are shown in Table 7.1. "Live/work" is when residential requirements take precedence over work needs and entail work uses with minimal noise, odour, and no employees or visiting clients. "Work/live" occurs when the needs of the work component take precedence, and there are noise, odours, employees, and visiting clients. While these activities could occur in the same unit, they could also occur in different parts of a single building or development. In this categorization, commercial live/work is the largest category of occupations and has the least environmental impact.

The range of live/work opportunities described in Table 7.1 outlines a framework for differentiating among the various types of home occupations and highlights the need for flexibility in zoning to allow a range of uses and opportunities in urban areas. Within an urban context, zoning that assesses and encourages compatible uses in a systematic and inclusive way is clearly required.

Implications for Neighbourhood Planning and Urban Form

Encouraging opportunities for low-intensity home occupations, neighbourhood telework centres, and support services in formerly single-use residential areas could potentially revitalize existing communities and provide an infrastructure for new forms of communities. Neighbourhood alternatives could alleviate the more negative aspects of working at home. Mokhtarian (1991b) foresees that the satellite or local work centre will ultimately be the most widely accepted form of telecommuting because, though initially more difficult and expensive to set up, potentially these centres have a broader appeal for both employers and telecommuters. This option combines the advantages of less commute time and regular contact with fellow workers, as well as providing a more professional image and reducing security and confidentiality concerns. However, subsequent research on several pilot telework centres in California has revealed that teleworkers do not like to use these centres and prefer to work at home (Mokhtarian et al. 1997).

Nevertheless, in different contexts these centres may be viable. For example, these work centres could be linked with an elementary school as an organizing principle for neighbourhood planning. Key services such as child care centres, copy shops with capabilities for desktop publishing, fax, and courier services, and cafés could be included, as well as recreational opportunities. One successful approach to incorporating commercial activities in residential areas are the executive office suites in North American suburban shopping centres, where small businesses can rent offices and sometimes support staff for a flat monthly fee. In addition, neighbourhood copy shops are becoming communication centres, where besides getting photocopies, a small business entrepreneur can rent computer time, use a laser printer, and fax documents or send them by courier.

Innovative planning and design projects are restructuring communities to integrate living and working in a pedestrian-oriented environment. "Pedestrian pockets" have been incorporated into the planning of suburban communities such as Laguna West outside of Sacramento (Calthorpe 1992). This concept balances and clusters jobs, housing, shopping, recreation, and child care. The integration of home and work life has been tried in some

Table 7.1

Live/work zoning typology

Category	Uses	Unit types
Commercial live/work	Office or service work with few or no impacts, no employees, no sales (e.g., self-employed consultants, researchers, software developers, analysts, writers, accountants, secretarial services; personal services such as hair stylists, tutors, doctors, therapists, child day-care; contract workers, teleworkers; office base for off-site services such as building and landscape contractors, sales reps)	Existing housing adapted for work or new housing that includes workspace/office
Commercial work/live	Above services , but where employees are involved, plus retail sales and repair or other services with frequent customer trade	Mixed-use building with a residential unit and workspace
Industrial live/work	Goods production or servicing involving lower impacts and no employees (e.g., some jewellery making, garment making, small leather goods, some printing, computer or small good repair, some production and recording studios)	Residential unit with garage, accessory building, or workspace with separate entrance
Industrial work/live	Goods production or servicing involving higher impacts, employees, and/or sales (e.g., metal work, wood work, some printing, some production studios)	Mixed-use building with residential, commercial, and industrial uses
Artist live/work	Artists and craftspeople working in low-impact media or processes (e.g., many painters, graphic, photography, and print artists; some potters, carvers; some musicians)	Renovated or purpose-built loft space
Artist work/live	Artists and craftspeople working in higher-impact media or processes (e.g., using amplified music, on-site film processing, welding, woodworking, spray painting, fired ceramics, generally using toxic or hazardous products)	Renovated or purpose-built loft space and separate workspace for toxic substances

Source: Adapted from City of Vancouver Planning Department (1996), 6, fig. 1.

cohousing projects built in Denmark, Sweden, and Holland where services, work areas, and some cooking are shared (McCamant and Durrett 1988).

Large employers such as Pacific Bell and the State of California are currently encouraging satellite offices in suburban locations that can bring together employees from various departments who live in the same geographical area (Nilles 1994). These employees are connected to their main offices via computers, modems, and fibre optic "throughways." Similar trials and programs are being implemented throughout North America (US Department of Transportation 1993).

Wiring neighbourhoods with fibre optic cable will allow the provision of a wide variety of home services including telework, teleshopping, teleschooling, and telebanking. Pre-wired broadband communications capabilities that permit the transfer of voice, video, text, and data will facilitate the connection of residences, schools, and businesses in a community through an Internet site and a community electronic bulletin board. This will allow greater information sharing and communication between community members on important local issues.

Montgomery Village, in a suburban community outside of Toronto, Ontario, is designed specifically for telecommuters and home-based businesses. Each residence is equipped with a home office that features access to Bell Canada's high-speed ISDN (integrated services digital network) telecommunications service; access to a local area network that will connect residents' computers to each other and to the local school and municipal office; and suitable layout – in many cases with a separate outside entrance to the office area.

While the integration of home and work activities could potentially create more livable, sustainable neighbourhoods, there is a real danger that a significant amount of home-based economic activity, especially telework, could stimulate urban sprawl and have other adverse impacts on land use and public transportation (US Department of Transportation 1993). People freed from the daily commute might choose to live further from urban areas, exacerbating the decline of the central cities, fuelling suburban growth and development pressures on rural areas. Correspondingly, this suburban and exurban growth could affect the overall pattern and density of land use in urban areas, impacting the design of public transportation. The limited data that was obtained from the Canada-wide survey did point to a pattern of exurban migration, and other studies have identified home-based work as a significant factor in the growth of these areas (Lessinger 1991).

As well, the reliance of Canadian home-based entrepreneurs on automobiles for their work raises questions about the effectiveness of home-based work activities in reducing automobile dependence. As was shown in the survey, most homeworkers still use their vehicles on a regular basis since there are few services and amenities in their immediate locale, and home-based entrepreneurs need their automobiles to visit clients and pick up and deliver goods. Home-based work is not a substitute for transportation for this segment of the work population. The planning of services and social and recreational amenities needs to recognize the presence of home-based workers in the neighbourhood, and orient local amenities and activities to encourage their use.

Though there are some problems with the introduction of home-based work into residential neighbourhoods, for communities as a whole (especially for those that have been adversely affected by economic restructuring and the subsequent loss of jobs in primary industries), home-based economic activities are a significant factor in stimulating economic growth (Shragge 1992). Home-based work allows small businesses to start and test the market before having to make large financial commitments. Reviews of policies and procedures in individual municipalities will assist in more fully recognizing this form of work as a viable economic activity. More awareness of the positive benefits provided by home-based employment will help counter public fears about change in character of neighbourhoods. Municipal officials, the public, and home-based entrepreneurs should be educated on the nature of telework and home-based employment and their significance.

While employed teleworkers seem to have the fewest problems adapting their home environment for work because generally they only work at home part time, the self-employed segment of the home-based work population were more mixed in this regard. Because their homes are their principal work sites (unlike teleworkers), they were more likely to require modifications to their homes to make them suitable for work. These include renovating or adding additional rooms. Many have considered moving to a larger home with a more appropriate layout.

In this chapter, alternatives were proposed to existing urban patterns that segregate and isolate functions and uses. Home-based work is one of many types of work that can be located in residential communities. Incorporating opportunities for work, such as telework centres and satellite offices, as part of a comprehensive strategy that supports the development of locally owned and operated enterprises encourages the regeneration of local resources and infrastructure. The local community and its resources are the starting point from which to formulate equitable and ecologically sound solutions that integrate workplaces, services, and homes.

It is yet to be determined whether telework and home-based employment will generate the return to widespread locally based community life envisioned by futurists. Nevertheless, telework centres and satellite offices may enhance the desirability of the neighbourhood as a locale for work-related services and as a respite from the intensity of home/work spaces. Other services, such as cafés, bookstores, and copy shops, may be generated by the use of these centres. In addition, the importance of the neighbourhood as sensory relief from the sterility of staring at a computer screen all day cannot be overlooked. The siting and landscaping of neighbourhoods to provide opportunities for pleasurable walking or biking take on added significance for homeworkers.

Home-based work may facilitate geographical dispersal, but it may also allow the promotion of local specialized communities based on common interests. People will have the possibility of conducting most of their production and consumption activities in their homes, but the need for face-to-face contacts will not be eliminated. Telework and home-based employment open up myriad possibilities for both weakening and reshaping communities and urban form, which will be explored further in the final section of this chapter. The following discussion explores interrelated concepts that need more careful examination now that working at home could potentially alter how homes and neighbourhoods function: community and sense of place. The evolution of these concepts reflects an emerging consciousness of the role each plays in "grounding" us to a place, and how that place may no longer be defined only in physical terms.

Meanings of Community

A community infrastructure that supports social interaction has been an implicit goal of planners and urban designers. These professionals assume that the designed physical order will facilitate certain kinds of social responses. Their approach, however, has invariably been physical, without any clear understanding of what community entails in modern society. It is a curious paradox that while designers still firmly believe that there is a traditional community to design for, social theorists would argue that community as a physically defined entity is being replaced in modern society with new forms of community.

The concept of community has always been one of the central themes of sociological theory, but it has also been the most difficult to define. "Community" is most commonly defined as "an aggregate of people who share a common interest in a particular locality" (Bender 1978, 5). In this definition, social organizations and social activities are territorially based in a locale, though the exact dimensions of this locale are difficult to determine. As a concept, community never stands alone but is always contrasted with noncommunal patterns of daily life. The German social theorist Ferdinand Tonnies developed a definition of community to explain the changes in social relations after the introduction of capitalism and the urbanization of society. He theorized ideal types to depict the process of social change: *Gemeinschaft* (community) was the "intimate, private and exclusive living together," and *Gesellschaft* (society) was the "artificial construction of an aggregate of human beings" (Tonnies [1887] 1963). Examples of *Gemeinschaft* were the family, kinship groups, friendship networks, and neighbourhood, while *Gesellschaft* was identified with the competitive and impersonal city. For Tonnies there was an inevitable evolution from a predominately *Gemeinschaft* pattern of social relations to a pattern dominated by *Gesellschaft*.

Using these European concepts, American sociologists have contrasted the idealized image of small-town America with urbanized American society. The American sociologist Robert Park, an early member of the Chicago School of sociology and a founder of urban sociology who studied small-town America, developed a definition of community based on "functional interdependence, group identification and loyalty, [in a] distinctive location" (Park 1952).

Instead of looking to small-town life for the definition of community, other sociologists have analyzed ways that community is understood in the urban setting. Richard Sennett (1970), an urban sociologist who studied American cities, described the search for fraternity and sharing in the 1960s as a search for community. He contrasted the myth of a purified community, a particular kind of social group in which people believe they share something, with a community that offers a disordered, unstable, direct social life where people are forced to interact with each other and develop a real communal sense, rather than one that is assumed but never tested.

Sociologists from the Chicago School have added further dimensions to the concept of community by describing it in normative terms through social interaction and social structure, and through the cultural and symbolic elements of community such as local customs, traditions, and symbols. Hunter (1974), writing on symbolic communities, contended that communities should also be considered as symbolic variables in a hierarchy of communities from which the individual can select. He viewed local communities as "collective representations" or symbolic "objects" of orientation, and as "situations" of action requiring definition by local residents through cognition, evaluation, and attachment.

Some sociologists question the theory of the evolution from a predominately *Gemeinschaft* pattern of social relations to one dominated by *Gesellschaft*. In his book on community and social change in America, Bender (1978) developed the argument for a bifurcated society where both forms of social patterns exist. In this definition community is based on human experience rather than the static terms of locality. Bender asserts that by accepting the fact that we will not find community among all our neighbours or fellow residents, we will find community "in our face-to-face and mutual relations that will enhance the quality of our public culture by freeing it from judgment on communal criteria" (150). This definition questions the understanding of community as a continuum from the communal behaviour of village life to the noncommunal behaviour of urban areas. Instead, it asserts that both communal and non-communal patterns of behaviour can coexist in the same locale.

Virtual Communities

Another layer of complexity has been added to the question of what constitutes a community with the use of telecommunications and information technologies to create social relationships that are not spatially bound. The television set and the telephone are the most important electronic technologies in the home, representing the two broad categories of home information technologies: the stand-alone or one-way receiving units and the communicating or networked units. With the integration of the telephone, television, and computer, all forms of information, images, and data can be easily transferred between locations, opening up the possibility for a diversity of communication flows that reflect various ways of producing and accessing information. These communication flows are modifying social relationships both within the home and between the home and the outside world.

Just as new forms of communication increase the potential for a variety of new social relationships, the corresponding potential is also present for problems with privacy, data security, and centralization of information. A science fiction novel, *The Shockwave Rider* (Brunner 1975) describes an Internet-type system that uses telephone lines to channel voice, data, text, video, and anything else that can be decoded digitally from any address in the world to any other. People move in and out of transient habitats and instant cities, their permanent addresses on the network only. Privacy has disappeared, as the global network can cross-check facts between agencies and each transaction is traceable through the network. People live in constant fear that some fact will affect their ability to find work or maintain a household. These people have no mediators between themselves, as individuals, and the global society. This is now approaching reality for many people. Increasingly, concerns are being raised about the ability of Internet hackers to track individuals and corporations and tamper with vital data. E-commerce companies can now track their users through "cookies" implanted in their Web sites so that customers can be targeted for future advertising.

Users of the Internet nevertheless believe that the communities they form on the network act as mediators for the more alienating aspects of the electronic age (Rheingold 1993). Mitchell (1995) contends that the Internet redefines our conceptions of gathering places, community, and urban life by operating under different rules from public places. Computer networks enable social worlds that transcend the limits of a specific place, widening the range of potential connections. Teleworkers who use these networks describe them as "virtual communities," electronic communities that fulfil the same needs as physically defined ones. They exchange information and ideas on a variety of subjects. Though people work alone, in their private nests, they still feel connected to the world.

While these connections are often personal, at the same time they are disembodied. By being able to control their interactions with others, people can avoid the unpredictable face-to-face exchanges that are part of a complex society. Frequent Internet users feel more comfortable interacting electronically because they often don't feel comfortable in more intense interactions. In addition, the physical communities where they live usually lack opportunities to interact in ways that are not part of the marketplace. While they open their homes to the global society, they isolate themselves from the world outside their doors. Heim (1998) sees this movement toward a technologically interlinked society as distracting from the real, physical world and as a force that could erode and destroy "community." He articulates the uniqueness of physical public spaces as "sharing that cannot be virtual because reality arises from the public spaces that people share physically – not the artificial configuration you choose but the spaces that fate allots, complete with the idiosyncrasies of local weather and a mixed bag of family, neighbours, and neighbourhoods. For many, the 'as-if community' [created on-line] lacks the rough interdependence of life shared" (42).

Home-based workers who are not active Internet users are even more isolated from informal public interactions. Their range of contacts is usually limited to intense interactions with family and friends, and impersonal interactions with business associates. While these are no different from an office worker's relationships, home-based workers lack a diverse range of support networks and rely more heavily on the support they do have. In turn, their relationships become more intense, with the corresponding problems that this situation entails. In addition, home-based workers tend to be more geographically limited – usually remaining in their homes to work, and only venturing out to buy supplies or go to work-related functions. They have no opportunities either in the virtual communities or in physically defined communities for informal encounters.

New forms of communities that are not spatially bound raise questions about the definition of community, now that face-to-face interactions need not be a component of that definition. Nevertheless, the quality of these communities may not be sustainable. By retreating from the complexity of the modern world into electronic space, Internet users are trying to control and refashion interactions, negating the variety of interactions and exchanges that are possible in the "real" world.

Interpretations of Sense of Place
In much recent environmental design discourse the concept of "place" has been replacing words such as "environment" and "building." Its emergence can be traced to the phenomenologists' interpretation of "authentic"

environments and the postmodernists' quest for contextual design. Placemaking research has focused on issues of environmental experience and meaning at a variety of scales. However, because place is "primarily an experiential category and intangible to the core" (Dovey 1985b, 94), it is difficult to define.

Humanist geographers emphasize that space is transformed into place through human involvement in localized activities and the accumulation of memories that bestow meaning and identity on a setting. Yi-Fu Tuan (1977) distinguishes between the attachments to places based on visual, imageable qualities, and instances where emotional ties are generated through prolonged exposure to and experience of places. Relph (1976), in contrast, distinguishes between an "authentic" and "inauthentic" sense of place. An authentic sense occurs when we feel we belong to "our place" both as an individual and as part of a larger community. An inauthentic attitude is evidenced when we have no sense of the significance of a particular place.

The main impetus for contextual placemaking in urban design has been the reaction to the utopian visions of the urban reformers of the nineteenth and twentieth centuries. Aiming to reorder the landscape of the industrial city, these reformers (Howard [1902] 1970; Le Corbusier 1929) decried the evil and decay of existing cities and proposed ideal urban forms that would create a good and just social order. Le Corbusier particularly focused on efficiency and the city as a machine rather than as a social/cultural network. As a reaction to the resulting orderly plans that were destroying the infrastructure of cities, critics (Lynch 1960; Jacobs 1961) began arguing for an experiential approach to planning based on user needs and first-hand observations of existing conditions. Cities, to these critics, were primarily arenas for activity, not just visual objects. Places were not discrete units, but rather physical, social, and economic continuities. More recent critics of idealized urban design settings (Krier 1979; Rowe and Koetter 1985; Rossi 1986) are trying to resolve these two seemingly conflicting urban design approaches to placemaking: the utopian quest for the ideal, and the experientially based ad hoc environmental decisions that constitute freedom of choice for a city's residents.

Sense of place has primarily been analyzed in terms of personal attachment to a place due to the presence of physical factors that symbolically resonate in our consciousness. It is a characteristic of a place that makes it memorable or distinctive, and has a high "imageability" (Lynch 1960). Particular cities and symbolic or sacred locations may have a strong sense of place that gives them a uniquely significant meaning for large numbers of people. Such places may be culturally specific, although some writers (e.g., Norberg-Schultz 1971, 1979) refer to natural places that have such a strong character (a "genius loci") that most people share the experience of, and sense of belonging to, them. People's consciousness of places that possess a particular significance may be intersubjective in the case of a community or neighbourhood, or personal in the feeling of being "at home" in one's own physical environment. Places where significant biographical events have occurred may also elicit a strong sense of place.

The sense of place, for the phenomenologists, involves the notion of being "at one" with the world. Contrasted to this is a sense of "placelessness" (Relph 1976) where one does not belong because of separateness from the meanings incorporated in the place. Lack of human scale and order in places contributes to a weak sense of place. Attachments to significant places are being eroded by environments that replace diversity with uniformity and are planned with efficiency as an end in itself.

The phenomenological interpretation of sense of place assumes all people have one coherent notion of a particular place. The implicit assumption of this compositional approach is that the world is understood through categories that are essentially unchanging properties and processes. In contrast, the contextual approach, as postulated by structurationists (Pred 1984), regards sense of place as very individually and socially constructed. It can only be understood in terms of the activities that people participate in daily. A sense of place is developed through the act of doing and through the relationships that direct those actions. It cannot only be described in terms of experience, as it is also a question of how one is socially situated and what kind of social practices one has access to. Rather than being singular, overlapping elements of sociability create different senses of place for the same locale. The concept requires sensitivity to the larger societal processes that are embedded in the everyday flow of events. The concrete interactions that make us feel at home or uncomfortable are complex, and can't be fully articulated because there are so many different perspectives to describe the phenomena. For any particular place, the question of who may participate in events and under what circumstances (i.e., who may or may not do what, when and where) becomes a critical component of the analysis.

Sense of Place/Sense of Placelessness
The experience for some teleworkers is one where they feel entirely "at home" neither in their own dwellings nor in the world outside. While their homes have a strong identity focus for them, at the same time home is tainted with all of the pressures of modern society. Their neighbourhoods, because they rarely use them, have few attachments. Their journeys are often limited to work-related activities, allowing little time to linger in special places. Those whose social relations are based on virtual communities are truly "out of place"; they have found a replacement for what is lacking in their physical environment by inhabiting another realm. The contemporary urban environment for teleworkers does lack sense of place, but rather than interpreting

this phenomenon as individually perceived, as the phenomenologists do, it must be understood as embedded in a social context. The fragmented character of a teleworker's participation in day-to-day activities (not unlike the general societal pattern) has weakened his or her attachment to people, places, and things. Particular locales still elicit felt attachments because of historical local and regional cultural variations, but the structure of teleworkers' daily lives negates strong attachments except to the home.

This, however, does not mean that teleworkers are "placeless," nor that electronically mediated interactions create the potential for "placelessness." This view of the information society ignores the integral relationship among places, technologies, and social relations. Places are not just settings for activities, they are also what historically has taken place in those settings. It is interesting to note that in 1989 the state of California designated "The Garage," the small one-car garage near Stanford University where William Hewlett and David Packard launched their electronics company in 1938, as a historical landmark and the birthplace of California's high-tech industry. It has become a major draw for high-tech tourists visiting Silicon Valley. Even in the world of cyberspace, places do elicit meaning.

Technologies on their own cannot alter the meaning of environments; they can only support the choices made in a particular setting. The process of becoming a place is linked with specific social and cultural forms of action, individuals' life paths, and transformations of nature. The sense of being "at home" is individually and socially constructed and can only be understood in terms of the activities that people participate in. Teleworkers' attachment to places varies according to their life experiences and the resources they have available. Economic class and gender (among other factors) affect those experiences and resources.

Communication Technologies and Urban Space

The use of telecommunications and information technologies both in the home and elsewhere can assist in the realization of new choices in living and working. Sweeping generalizations, however, cannot be made about different technologies. The existence and use of a technology alters material and social givens, creating new options for, and new constraints on, individual action. Each technology is used differently and has distinct consequences. The same technology may be used by different people in different ways to different effect.

Despite this complexity, several distinct patterns are visible in the relationship between these technologies and urban space. In 1932 Frank Lloyd Wright, describing a new urban form for North America, wrote about two major inventions: the motor car and electrical communication in the form of the radio, telephone, and telegraph. These inventions, he contended, made centralization unnecessary and immeasurably widened the areas of man-movement. In his description of Broadacre City, his model for contemporary suburbia and exurbia, the city is not an arrangement of roads, buildings, and spaces. It is a process, rather than a form, made possible by mass acceptance of transportation and communications technologies. Broadacre City is a vision of the whole nation urbanized. Laid out on a grid where main roads are one mile apart, the gross density is two acres of land to each family. Industry and commerce are disposed along freeways. With its homes dispersed and disappearing into the landscape, and its mixture of agricultural land and industrial plants, this new city pointed to the end of concentrated urban development.

Wright borrowed his ideas heavily from an important study on communication agencies and social life published by Willey and Rice in 1933. Their study identified two contradictory tendencies. Communication and transportation technologies reinforced community patterns of attitude and behaviour, as contacts with others in the same community multiplied through increased local mail and telephone calls, local mobility of the automobile, and the fostering of local spirit through local newspapers. At the same time localism was undermined as the dissemination of information – through films, radio broadcasting, national advertising, increased travel, and long distance wire and wireless facilities – contributed to standardization over wider national and international regions. The authors hypothesized that the increase in overt standardization might be accompanied by retention of inward differences. Though the communication agencies are part of a complex interrelated network that fosters the existence and continuity of social life, some generalizations can be made. The transportation system and point-to-point communications tend toward localization of communal life, while mass communications tend toward standardization.

De Sola Pool (1977) and others have advanced the theory that the telephone contributed to the decentralization of urban spaces and the altering of social and spatial relationships in cities. The telephone aided urban sprawl and mass migration to suburbia by allowing communication without face-to-face contact. By enabling dispersion of services, the telephone promoted the process of neighbourhood decomposition, from a walking city to a vehicled one, and played a part in the imposition of zoning. It also helped realize the physical possibility of developing communities without contiguity. By means of the telephone (and now the computer connected to the telephone), information is easily available without physical proximity. However, while dispersion is one consequence of telecommunications technologies, they may also contribute to the creation of enclaves based on economic class and lifestyle for those who don't need physical proximity to their work but want security and face-to-face contacts with like-minded people.

Theorists are now speculating that the integration of the computer, telephone, and television has reinforced the pattern of dispersion of services across a wide area, and the concentration of specialized services and like-minded people in particular areas. It has been argued that a new economic order is being created with spatial consequences: a new "space of production" is being generated both by the importance of information in producing, utilizing, and transmitting products, and by the capacity of firms to separate and disperse internal functions to a variety of locations (Castells 1985, 1989). At the same time, the use of telecommunications and information technologies tends to reinforce the trend toward the concentration of corporate headquarters and support services in a few "world class" cities that are linked to one another as the dominant producers and users of information. In this way a dual order is created – the world cities and the rest – fostering a new urban hierarchy of cities (Friedmann and Wolff 1982; Moss 1987; Sassen 1991; Graham and Marvin 1996). For example, while e-commerce ventures are starting up all over the world, the heaviest concentration is still in a few select locales – the Silicon Valley outside of San Francisco, New York, and London – and businesses that want to go global have to have a presence there. Locational proximity is still important in the new "cyber-economy."

Many cities are now oriented toward a strategy based on the wide variety of uses of telecommunications. In Japan, a redesign is being attempted that will revitalize Kawasaki City, an industrial centre between Tokyo and Yokohama, and carry it into the twenty-first century as an information-intensive and humanistic city. Social and cultural development will be given at least as high a priority as economic elements of the design, whose stated aim is "to develop information and communication resources as the means of transition from the twentieth century, industrialized, product-intensive city to a high technology, knowledge-based city of the 21st Century" (Japan Association for Planning Administration 1986, 10). A new Malaysian city based on the same principles, Cyberjaya, is emerging from the jungles near Kuala Lumpur, though there is considerable uncertainty about its viability.

The telecommunications infrastructure for the new information city evolving around the world includes a phone system with digital switching, miles of fibre optic cable, universal connection to an interactive cable television system with adequate institutional channels, numerous local area networks (LANs) as well as satellite linkages, and locally available value-added networks for general information or specific business functions. In this new city, office buildings are being developed with "shared tenant services," and "smart buildings" are providing the answer to the problem of accessing telecommunications systems for medium-sized and small firms. This telecommunications infrastructure has ramifications for the internal spatial organization of cities, affecting what types of land uses and developments can occur, and the distribution and use of services and space at the neighbourhood level.

The neighbourhood – as a basic unit for city building and functioning, with an identifiable locale and an infrastructure of services and activities – is also changing. Services associated with the neighbourhood, such as local bank branches and theatres, are being replaced by automated tellers and the diverse services provided by the telephone, television, and computer. Services in the home such as telebanking and teleshopping, cable television and direct broadcast satellite (DBS) systems for entertainment, the replacement of newspapers with Web information sites, remote home utility meter monitoring, electronic mail, telemedicine, and telework could result in the internalization of much public life in the home, dissociating it from the neighbourhood.

While the dispersion of services at the neighbourhood level has not been empirically substantiated, the use of telecommunications and information technologies does appear involved in the trend toward segregated residential communities. These enclaves with a defined spatial identity, where people of the same economic class and lifestyle congregate, have often formed to protect their residents from feared elements of society. Now that work can be dissociated from a particular setting, more people can choose where they want to live based on other criteria than the journey to work. While the segregation of residential communities was occurring long before the significant use of these technologies, nevertheless, the trend of exclusive communities outside of large metropolitan areas, miles from work opportunities, is prevalent in the dispersion of residences across a wide geographical area and the reconcentration of residential settings in enclosed compounds protected by security personnel. A 1997 study (Blakeley and Snyder) found that there were 20,000 of these gated communities in the United States with more than 3 million housing units. An even more telling statistic is that 54 percent of home shoppers (i.e., teleshoppers) surveyed in southern California wanted to live in gated and walled developments. The authors theorize that the gated community can be seen as reflective of North American society and culture and "is a dramatic manifestation of a new fortress mentality growing in America" (1).

Just as the futurists have been overly optimistic about the positive effects of technological change on home life, their critics may be overly pessimistic about the consequences on the use of public space of the widespread introduction of telecommunications technologies into the home. While no conclusive research indicates that the use of communities and neighbourhoods intensifies when people work at home, there are also no indications that the use of public open space is decreasing or that community activism is on the decline. While new information technologies may be creating a "placeless" society, where geographical location will have no real significance, this has not been empirically confirmed. There are also indications that not all of the electronic services that can be provided in the home, and that can potentially substitute for community activities, are widely used, though accounts in the popular press predict that Internet transactions will increasingly become the norm. Nevertheless, the floundering profits of many Web-based e-commerce businesses attest to their lack of general acceptance. Consumers do, however, perceive information technologies as a viable substitute for interactions in physical space in terms of saving time and as a convenient and instantaneous way of gathering information.

What telecommunications technologies in the home do allow is the creation of social worlds that transcend the limits of place. It has been suggested that the telephone created "psychological neighbourhoods" of physically dispersed social relationships, sustained the extended family, and provided a sense of security and sociability (Fischer 1985). Similarly, special-interest Internet discussion groups connect people worldwide. For intensive computer users connected to a range of listservs, interactions over the computer, from exchanging information to meeting partners, are substituting for face-to-face socializing. Communities based on common interests rather than physical locale are now possible, even though it is not clear whether these communities are replacing the friendship patterns and social relations found in physically defined communities.

The belief that the most authentic personal community is that which is the most local is now being contrasted to the theory that "modern life allows people ... to create social worlds almost entirely free from the artificial limits of place" (Fischer 1982, 158). Meyrowitz (1985), in his description of the impact of electronic media on social behaviour, further asserts that the social meaning of settings is no longer shaped by place-specific locations. He contends, "To be physically alone with someone is no longer necessarily to be socially alone with them ... by altering the informational characteristics of place, electronic media reshape social situations and social identities" (117).

The Telecommunicated City

Castells (1985, 1989) has argued that the trend toward the "telecommunicated city" has increased the functional zoning of time and space, creating a concentration of activities around workplaces, homes, and leisure. The distribution of people and activities that land-use zoning has tried to regulate is further segregated by "electronic zoning." In other words, the city is evolving into a dual society of haves and have-nots based on access to telecommunications and information technologies. In this society, the middle-class home is becoming increasingly atomized and individualistic. Ethnic and low-income communities are being excluded from most information networks, except commercial television, which is becoming increasingly more restricted in its coverage. These communities consume inexpensive television and rely more on direct interaction in the community.

The new elite are the information-rich (or overloaded) who are taking over the desirable central spaces of the urban areas. While these new urban patterns are difficult to corroborate empirically, the time/space profiles developed in the previous chapters do chronicle a new relationship that teleworkers have to their cities. Their spatial reach becomes largely limited to work-related journeys, with the majority of their time spent in their homes.

The research also found that home-based work does not necessarily make the neighbourhood more important as a social and service locale for teleworkers. In North American society the need for social ties and a sense of belonging is now met, in many cases, by the workplace. When those ties are severed, individuals must develop new forms of community. For some teleworkers the neighbourhood can fill this need by providing opportunities for informal encounters. But for most, the neighbourhood offers few attachments, since teleworkers have few social contacts in their neighbourhoods and do not participate in many activities in their immediate communities.

This chapter has speculated on the transformations that are and will be occurring as workplaces decentralize and homes and communities adapt to new uses. For many, the workplace has become an inefficient and unstimulating environment for creative interaction. Companies, as part of corporate restructuring or "reengineering," are downsizing corporate headquarters. People are now working almost anywhere: at home, in clients' offices, even in their cars. In the future a completely decentralized work organization is envisioned with no need for a central corporate entity or facility. People will form "virtual companies," fluid organizations that will be set up for specific projects and then disbanded after their completion.

However, the implications of work reorganization for home and work life have yet to be fully articulated. Information technologies allow opportunities for both freedom and social control. While it is also envisioned that "virtual communities," electronic networks, will replace the need for physically defined communities, the preceding pages have explored the difficulty of defining community in this context. Electronic networks enable us to communicate with anybody, anytime, but that does not mean that we will want to.

Among the consequences of new forms of work organization in the urban sphere are the dispersion of services across a wide geographical area reinforced by the integration of the computer, telephone, and television, and the concentration of specialized services and like-minded people in particular areas. In addition, the trend toward the "telecommunicated city" is intensifying the expansion of the private domestic sphere and the corresponding abandonment of uses and functions in the public sphere. Coupled with this concentration of activities in the private sphere is the potential for an increasing polarization between the private and public. As our private world broadens, our public world becomes more remote and impersonal. As a result, our public spaces become anonymous, fearful places, while our private space reinforces the tendency toward narcissistic autonomy. The private/public split is no longer my space and our space, but my space and nobody's.

8
Conclusion

Given the pervasiveness of technology in every aspect of our lives, there seems to be surprisingly little critical discourse on how we want our personal and work lives organized in relation to new technologies. There seems to be a lack of self-reflection on the power of technology to transform our lives, societal amnesia regarding its consequences, and little discussion of how things could be done differently.

While recognizing my relatively privileged position and the financial resources I have available, I am struck by how much I embody my research. The demands of my job have increased significantly in the last few years as my university department copes with early retirements, fewer faculty, staff, and financial resources, and greater expectations of productivity. Besides working at my office during the day, most evenings and weekends I turn on my home computer and work as well. I always check my e-mail early in the morning before going to my office, otherwise I may not get to it for the rest of the day. Often, I have to get up by 5:00 a.m. to do this and have some time for research; if I awake later, missed tasks may take days to get back to. In the later stages of finishing this book I was getting up at 3:00 a.m. During the day the demands of my job – my teaching, advising, and administrative duties – rarely give me time for quiet reflection. I rarely rely on support staff to assist me in my work.

I know the conflicts in trying to combine work and family life. I am always distracted at home. My five-year-old daughter often tries to grab my keyboard to get my attention. Aside from focusing on her needs, there is always pressing work that needs to be done. I do not have a separate workspace in the home, but share my space with the television and my daughter's play area. I know the unresolved tensions of wanting my own space but also wanting to be close to family activities.

I try to have at least one day during the workweek when I am at home. Like many other academics, my productive writing is always done at home, when I am alone, after I have taken my daughter to school. Working at home allows me the uninterrupted time I need. Because I am connected to my office by a modem, I could be working anywhere that allows me instant communication. My office I see as a place for "busy work," advising students, teaching, attending meetings, and handling administrative matters, not for creative work. However, I can appreciate why the respondents in Hochschild's study (1997) derive so much satisfaction from their work environment as opposed to their home life. I place a high premium on my work even when it is conducted at home. My paid work life gives me status and financial rewards. I can measure outcomes by products and feel accomplished when problems are solved. My reproductive role is not nearly as easily evaluated. It is repetitive and messy.

An acquaintance is even more consumed by his work. Whenever we visit him and his family he is always on his laptop computer, seemingly using it everywhere in the house. His wife estimates that he is sometimes on-line ten to twelve hours at a stretch. An academic in the IT field, he boasts that he created his reputation and got his current job entirely through contributions he made on various on-line discussion groups. He does no household maintenance. His wife and two teenage children beg for his attention, but he is too distracted to engage with them. His daughter says that she cringes when she hears the sound of her father's computer modem dialling up to the Internet.

I observe myself being irritated by those who do not seem to have a commitment to their work similar to mine (i.e., who don't work on weekends and after traditional work hours), and I am apprehensive of those I term "technologically challenged." Another acquaintance (a male in his fifties who is head of his department) acknowledges that he is barely computer literate, seemingly unable to learn essential skills such as sending documents by e-mail. While he is profusely apologetic for the inconvenience caused when collaborators have to re-input his material, I can only interpret this inability to learn basic computer skills as a convoluted way of maintaining institutional power and resisting the process of dis-intermediation. He has relied on his secretary for computer support, but the workload of university support staff has increased significantly due to fewer staff. His secretary recently took extended stress leave, and he is now on his seventh temporary assistant.

As a woman in a largely male-dominated profession, my perception is that I do not have the luxury of a gatekeeper, such as my colleague's secretary, to intervene between the technology and myself. I also seem unable to set work boundaries that could clearly define my work and private time. Consequently my expectations of my workload have grown. Because I know that I can do it, I don't question whether I should take on new work, even when it will obviously require even more hours disengaged from my domestic life. My resistance strategy to the pressures of my work is not to clearly define boundaries but to opt to work in my home whenever I can. While recognizing the toll this takes on family life, I also know that without the ability to work almost anywhere I would not be able to maintain both my productive and reproductive life. I have become adept at using technology to navigate my life.

Given that these experiences are probably being replicated many times over in other households, what are the consequences for the social fabric of society? Postman (1992), a communications theorist, stresses a fundamental question that we should all be asking ourselves: "Am I using the technology, or is the technology using me?" I know how consumed I have become by my work and how the capabilities of the technologies have aided that relationship. I am wired into my work on a continuous basis. I think that I can shut it off, but in my heart I know I cannot – I am seduced by what the technologies offer. Disengaging seems too stressful and a risky proposition given the mounting expectations of the corporate entity I work for, the university. My home is tainted with work, and though at times the power that the technology affords me is exhilarating, it comes at a price.

In the course of this research I have attempted to understand the relationship between technological change, social change, and spatial structure. While because of the many extraneous factors that may impinge on these variables this research looked at tendencies rather than direct effects, clear patterns in daily activities have emerged. Technology is not a determinant but an enabler of social change. It is not neutral and value-free. Technologies are socially constructed, and the processes through which they are developed reflect imprinted social biases. The effects of technology do not reflect only the circumstances of its use, but larger macro processes that are manifested in the micro scale.

As a society we seem unable to disengage enough from new technologies to critically understand their impact. Critiques are often simplistic – either unrelentingly upbeat (Negroponte 1995) or fraught with doom and Ludditism (Sale 1980). Rather than merely championing women's access to technology or the use of technology for activist purposes, we need a feminist critique that revisits fundamental questions regarding technology and its impact on the experiences of daily life. Computers and their uses, especially the prevalence of the World Wide Web, usurp increasingly more of our time, fitting into the structure of increased expectations. While no longer necessarily seen as a threat to jobs, as in the 1980s, technology is now seen too uncritically as the saviour of employment. As economic and social globalization has overtaken our lives, we communicate electronically with people from around the world, but rarely with our next-door neighbours.

Telework is a new form of work resulting from technological, economic, and social change. While this new form of work has allowed for greater flexibility and control in how, where, and when work can be conducted, tensions are manifest between the desire for flexibility (by both corporations and employees) and the tendency for work to invade every aspect of one's life. While this phenomenon must be considered against the background of other changes that are occurring in society, significant conclusions can be drawn from the home-based work experience to describe new societal patterns of home and work life. While many home-based workers appear to benefit from working at home, homework is not a seamless utopian intermingling of work and domestic responsibilities. Working at home may intensify existing problems, such as difficulty in separating work and domestic activities. As well, home-based workers may be prone to feeling isolated.

A more comprehensive understanding of the implications of telework and home-based employment for policies at a variety of levels is clearly warranted. While home-based work may be beneficial for some home-based workers, affording them flexibility and control over their time and space resources, for the majority, especially low-income women, working at home is a survival strategy that affords them few opportunities for advancement. Those for whom it is a positive experience find they are generally satisfied with their work, their working conditions, and the opportunity for work to be combined with other activities. Those who are not satisfied find that home-based work results in role conflicts, inadequate workspaces, the blurring of the division between work and domestic life, and an increased tendency toward "overwork." Home-based work is not a panacea for unresolved tensions in the work and domestic spheres.

Home-based work is regarded as a potential instrument for improving working conditions and the general quality of working life, as well as contributing to the sustainability of communities. Socially it has implications for commuting patterns, the transportation system, child care services, and a variety of work-related resources. If a significant portion of the population were encouraged to telework, it could have a major effect on the decentralization of corporations. While telework has the potential to contribute to more livable and sustainable communities, the evidence is not in regarding its actual impact. Data from the studies in this book support neither the reinvigoration of communities nor the democratization of corporations.

Cautions need to be voiced about the impact of this type of work on our social and physical infrastructure and on organizational structures. Many critical issues are not being addressed as the workplace profoundly changes. Does contracting out work to at-home workers lead to exploitation and problems in combining family life and paid work? Does home-based economic activity encourage an individualistic and atomized society? Is a "footloose" labour force increasing urban sprawl, now that residential choices need not be predicated on the journey to work? These are strategic and ethical dilemmas that policy makers need to address as they try to both encourage the beneficial effects of at-home work and prevent its exploitative or destructive aspects.

The corporate restructuring that is occurring under the term "reengineering" is leaving the workforce increasingly atomized and stratified. Fewer employees are expected to do more with less supervision and support. There are inherent tensions between corporate efficiency and workers' desire for flexibility and control, resulting in the loss of jobs, security, and the work environment as a community. Privacy issues are also not being fully addressed as the home becomes a workspace. For teleworkers and home-based entrepreneurs, the home is losing its nature as a refuge as work-related stresses become associated with it. Effectively, teleworkers are wired into the job twenty-four hours a day and can be monitored electronically at any time.

The home in Western culture has psychological and sociocultural significance as a refuge from the rest of society and an extension of the self. In an information society where knowledge itself is a key source of power and innovation, large systems and institutions appear to many people far beyond their understanding and control. Public life is seen as the arena of experts, distant decision makers, or simply the rich and famous. Public space is often feared because of the unknown elements that lurk there. The home becomes, for many, the only venue where they can feel a sense of control over their environment. The meaning of home emerges from an interaction of opposites: public versus private realms, other versus self. For those home-based workers who are locked away in their home spaces, labouring long hours over their work, rarely venturing out, the home takes on an added dimension. For them, the home becomes both comforting and confining. With the introduction of paid work into the home, new patterns have emerged reflecting the rhythms and priorities of the corporate culture. Correspondingly, the home is no longer a buffer against the external pressures of the market place. The postindustrial society is "coming home."

In addition to the loss of the home as a buffer, teleworkers' relationship to their physical and social communities changes. Besides the lack of office socializing, homeworkers no longer have the area around the workplace as a place for recreation and meeting friends. Office workers often use the journey to work as a multipurpose trip – they shop and socialize after work. Teleworkers use their larger community setting less, without substantially increasing neighbourhood use. The neighbourhood is rarely used as an outlet for informal socializing as homeworkers don't allow time for this, nor are most of the neighbourhoods they live in conducive for neighbouring activities.

The shift of paid work into the home is precipitating the development of a new social identity, a blurring of domestic and work associations. In turn, the home is taking on new social meaning. In some situations homeworkers benefit from greater flexibility and the time saved by not commuting, but for most the constraints are exacerbated. Their social connections are limited and they are "chained" to their computer terminals. More time is spent in the home, alone. Face-to-face contacts are replaced with other forms of communication. This does not necessarily reduce contacts, but it formalizes and depersonalizes them.

While many homeworkers have autonomy, few have a wide spectrum of significant connections that nourish and sustain them. For those in family settings, the reliance on intimate family relationships for their sole emotional sustenance can become too intense. For those who are single, emotional life can become impoverished. Working at home often intensifies existing conditions; it does not resolve them. For example, working at home does not change the division of labour within the home. There is no fundamental shift in attitudes or practices, although men do slightly more housework and child care when they are at home. Nevertheless, men still see themselves as primarily working, even at home; women are still torn between their work and family responsibilities.

North American society has emphasized individual initiative within a conformist consumer society, creating the myth of the rugged individual pitting himself or herself against society. Teleworkers perceive themselves in those terms: they are "homesteaders" exploring the new frontier of home-based electronic work. The social cost they have to pay for their freedom, however, is that they are alone in that new terrain. Homeworkers are especially prone to feelings of isolation. Because of how they work and live they have few opportunities to develop community connections. While isolation is a difficult condition to resolve, not finding healthy ways of alleviating these feelings can result in unhealthy behaviour, such as secret drinking.

The social polarization resulting from bifurcated labour markets in North America is creating a highly skilled pool of mobile informational workers, the majority of whom are men, and a vulnerable low-skilled pool of predominately women workers, who are often visible minorities. This is simultaneous with the disappearance of social struggles that might influence the state to serve interests other than that of capital (Castells 1989). The garment industry, with its shift from factory to industrial homework, and the low-income communities of women, youth, ethnic minorities, and immigrants from which these workers are drawn, exemplify this bottom sector of the urban economy that is being forgotten in the new political life of the investment-chasing entrepreneurial city. Data entry and call centre workers, largely drawn from a similar labour pool, are a new form of sweatshop worker. Highly skilled telework, in contrast, is part of the "space of the new upper tier ... connected to global communication and vast networks of exchange" (227).

While telework outwardly appears to serve the interests of the corporate elite who seek flexibility and control over their working lives, these elite workers may be just as susceptible to "invisibility" if their change in work venue is followed by a change in work status. In corporate telecommuting programs that attempt to decentralize organizations, there is a real danger that home-based workers will become marginalized and therefore susceptible to reclassification as independent contractors or pieceworkers. Trade unions fear that employees will lose their benefits and bargaining power with these new working arrangements.

Home-based workers, in general, believe that they are more efficient and productive, and research on corporate productivity corroborates these feelings. Specifically, telework improves productivity, cuts operating costs, allows additional pools of labour to be tapped and utilized, achieves flexibility in personnel management, and cuts social costs (Korte, Robinson, and Steinle 1988). These findings, however, reflect the experiences of more skilled workers and professionals.

For low-skilled workers, such as data processors, the degree of autonomy remains low; their status does not change because of a change in work venue. As recent studies illustrate, low-skilled workers are the most vulnerable and the most in need of protection (Pearson and Mitter 1993). The disparate nature of contract work and piecework, coupled with domestic obligations for this predominately female population, renders it virtually impossible for workers to organize themselves effectively, making them vulnerable to exploitation and willing to accept substandard working conditions. These workers, especially at-home independent contractors, need to be recognized within corporate structures and given insurance and statutory protection similar to that given employees in conventional workplaces. In the Scandinavian concept of participatory design first introduced in the 1980s, workers and management redesign workplace technologies together, to reflect democratic ideals and preserve skill content of jobs. This and other models developed elsewhere need to be investigated for their relevance to the Canadian context.

Strategies for Intervention
The preceding discussion has focused on telework's potential as well as the vulnerability of teleworkers (especially independent contractors and back office workers) to exploitation due to their invisibility and lack of opportunities for collective action. What can be learned from this discussion? The main issue surrounding telework is unequal access to resources, infrastructure, and training. While women in Canada see the economic opportunities that access to information and communication technology can provide, they still lag behind men in the use of new technologies and applications. Training in Canada is inadequate to keep up with the demand for skilled ICT workers, and in addition it is hard to keep trained women workers with families, because of the difficulties of combining child care with demanding

careers. In developing countries, the establishment of the infrastructure and training for ICT careers is further impeded by national and individual poverty. Coupled with the need for training is the need for recognition that child care is a societal issue and must be addressed at that level.

In Canada there is a growing disparity between the urban and rural areas, as poor infrastructure for telecommunications is impeding rural development opportunities. However, people in remote and rural communities understand that access to on-line services will provide them infinite possibilities in education, entertainment, networking, health, employment opportunities, and a market for their commerce. For example in Nunavut Territory, the largest Internet service provider has achieved a penetration rate of 1,400 subscribers in a population of just 5,000 (Boei 1999). On a micro level there is an emerging need for communities to optimize the opportunities that information and communication technologies present. In particular, individuals and groups should be supported in using the technologies to overcome disadvantages such as physical isolation, lack of skills, or poverty. "Community informatics" is an approach that links economic and social development at the community level to ICT opportunities such as Internet and electronic commerce, flexible networks, and telework (Gurstein 2000).

The home-based work population has multiple socioeconomic constraints linked to its position in the labour market. A 1995 International Labour Organization report on homework states succinctly, "Homeworkers constitute a particularly vulnerable category of workers due to their inadequate legal protection, their isolation and their weak bargaining position" (1995, 1). These findings, mainly related to industrial homeworkers, and substantiated by similar findings in a variety of contexts, raise important issues pertaining to socioeconomic inequities and low employment standards especially for low-income female workers. In response, various jurisdictions have identified a number of strategies to address these problems. While these strategies target the most vulnerable population of home-based workers, they contain lessons for the entire population, given the common stresses and problems facing home-based workers.

One strategy has focused on legislative action aimed at employment standards provisions and the regulation of employers of independent contractors. This has included regulatory measures to ensure that homeworkers are protected by employment standards legislation, including assurance that industrial homework is appropriately defined in legislation and policy, appropriate minimum wage and remuneration laws are in effect, the terms and conditions of work are regulated, social security provisions are available, and occupational health and safety conditions are regulated. To reinforce this regulatory framework there have been efforts to establish reporting, registration, and supervision systems for industrial homeworkers (Ontario

District Council 1993; Women and Work 1993; ILO 1990). However, regulations have only a limited impact because employers evade them and they cannot be effectively enforced, due to the clandestine nature of homework. Also, the material conditions of homework mean there are few opportunities for collective bargaining (Mitter 1992). In response to the difficulties in enforcing legislation, local initiatives have concentrated on social action and community economic development measures that empower and support homeworkers through research on their conditions, organizing and advocacy, and the development of homeworker centres (Rowbotham and Mitter 1994).

In a very different socioeconomic context, the Self-Employed Women's Association (SEWA) in India is one of the most successful examples of organizing homeworkers (ILO 1995). Its membership of approximately 40,000 consists of small-scale vendors, home-based producers, and labourers selling their services or labour. Of those, home-based producers (such as cigarette makers) were found to be the most exploited and lowest paid (62). Since gaining official recognition as a trade union in 1972 by arguing that though its members are considered self-employed the priority of the union was the unity of the members, SEWA has ensured that homeworkers are covered by the Minimum Wages Act, has conducted training courses to upgrade skills, has organized cooperatives for production and marketing, and has created savings and credit cooperatives to provide loans. Internationally, SEWA is considered a hybrid organization, "part union, part cooperative, part bank, and often considered by the development community as an NGO" (Prügl and Tinker 1997, 147).

The principal community economic development vehicle for industrial homework in several jurisdictions has been the Industrial Homeworker Centre (Rowbotham 1993). These centres have been established formally in different cities in the United Kingdom and Holland and informally in many other developing and industrialized countries. A homeworker centre is a local neighbourhood centre that provides an alternative setting for work previously done at home and meets other needs such as child care, skills upgrading, language training, drop-in social facilities, merchandising channels, and other services determined by user needs. It could be very small-scale (e.g., a small cooperative initiative out of someone's home) or it could be larger (e.g., at the level of a women's centre). Neighbourhood telework centres as described in preceding chapters are a similar alternative to working at home for teleworkers and home-based entrepreneurs.

Industrial homeworker centres go further than the telework centres being developed in North America, as they support women in community economic development initiatives to improve the quality of their lives and communities. Such initiatives could potentially result in provision of language training and child care, improved health and safety conditions, at least within the homeworker centre, and networking and training opportunities. In turn, the collective support found in these centres could lead to forms of collective bargaining, a better trained labour pool, and increased economic returns to industrial homeworkers. An example of this is the formation of small sewing cooperatives or artisan groups of women who receive orders higher up the production chain and therefore earn more of the profits from their assembly work. This has been tried with varying degrees of success in Italy and Britain as well as many countries in the South (Totterhill 1992). Telecottages developed in rural areas of the United Kingdom offering services and training, described in Chapter 2, exemplify how this idea has been interpreted for teleworkers.

Besides community measures that address the material conditions of homeworkers, microcredit is an increasingly popular tool for poverty alleviation and social development around the world. It especially targets self-employed entrepreneurs, many of whom are home-based and female. Microcredit can be an especially powerful tool in providing start-up funds so that low-income women caught in the cycle of piecework can develop enterprises of their own. It is estimated that nearly eight million people in developing countries are currently beneficiaries of microcredit (Micro-Credit Summit Draft Declaration 1996, 6). Microcredit programs are part of a movement toward alternative development methods, which target people at the grassroots, provide local control of development processes, and take gender into consideration. However, without fair working conditions, such as the minimum wage standards negotiated by SEWA, homeworkers who are recipients of credit programs are still susceptible to below-subsistence wages (Prügl 1996). To address these problems, the two strategies of union organizing and development of home-based entrepreneurship should be combined (Prügl and Tinker 1997).

Given the vastly different material conditions in the North and South in the illustrations above, and the seeming reluctance in the sample I surveyed to engage in any form of collective action, what interventions are possible? Associations of at-home workers need to be strengthened in their mandate and their appeal. Home-based entrepreneurs have several organizations that they can join to get information and lobby for government changes regarding their status. However, independent contractors rarely know people doing similar work that they can go to for support and assistance. Union representation would ensure that these workers would at least have minimum working standards, but because of their invisibility and the opposition to unions by employers, union organizing has only had a limited success. Government recognition of the rights of these workers would have to be in place before their working conditions could improve.

Alternatives to working at home need to be explored as vehicles for community economic development. Given the benefits of locally based small-scale economic activity, various levels of government and the private sector need to partner to recognize and encourage these activities through incentives. The high-skilled and the low-skilled IT independent contractors need to form coalitions. The emphasis in the new economy on the atomization of workers negates a common understanding of the material conditions of contract work. These issues need to be assessed across the broad spectrum of worker experiences. While the World Wide Web is predominantly being used to create a global marketplace, it is also being used to create communities and solidarity groups. Low-skilled IT workers need to tap into these networks.

Concluding Comments

Home-based work is not a return to an utopian time when family and work responsibilities were intermingled. Historically, that idyllic life existed for only a very few. For the rest, work based at home meant constant work for every member of the family, with little free time. This is also the experience for most present-day homeworkers; women in particular rarely have leisure time. Work is spread out over most of the day, resulting in less time for other activities. This raises the issue of what "flexibility" really entails. While telework appears to increase productivity and in some circumstances allows work to be combined with other activities, it also results in role conflicts, inadequate workspaces, the blurring of the work/leisure time division, and an increased tendency for "overwork." The home can be unsuitable as a workplace for many people because of spatial constraints and the lack of social contacts. Homework inhibits face-to-face interactions, resulting in social isolation. Coupled with isolation is the feeling of being "invisible" to fellow workers, friends, and family who don't perceive teleworkers as really working. Home-based employees feel disassociated from the corporate culture and their opportunities for advancement are curtailed. Many employees are not self-motivators and cannot cope with managing their home and work responsibilities in the same environment.

Working at home may not be the only alternative. However, the flexibility and control that seem to be the prime motivating factors for home-based workers are difficult organizational principles to accommodate within many existing corporate structures. While the technology is available to allow various alternatives, the major hurdle is the willingness of employers to restructure corporations to create less hierarchical organizations that could accommodate more flexible work arrangements. Managers are threatened by the lack of control they have over workers who organize their own work schedules and venues. New ways of working are only possible if corporations recognize the importance of workers having ownership of their work. Given the reluctance of many corporations to change their corporate cultures, it is unlikely that formal telework programs will become a widespread phenomenon.

While in many ways the teleworker experience is not particularly positive, it is preferable to other alternatives because it affords the freedom to organize time and space in ways that suit workers and their families. If telework is organized with the needs of workers and their families in mind, it has the potential to allow greater flexibility in daily life. But we need to go further than that. With social policies that allow more balance and options in the work/family equation, and an integrated approach to the planning of homes, workplaces, and communities, all people, but especially women, could participate in both domestic and work activities. The challenge is to develop telework policies that encourage social networks and enhance the range of choices for workers as well as their employers. If telework is to be a viable alternative, the consequences of a solitary work life need to be explored and addressed as a societal issue.

Appendix A
California Study on the Social and Environmental Impact of Working at Home

Part 1

Profile of Respondent (All)

1 Sex
   1 Male
   2 Female

2 Type of worker
   1 Full-time homeworker
   2 Part-time homeworker (four days at home, one day at office)
   3 Part-time homeworker (less than four days)
   4 Office worker

3 Employment status
   1 Self-employed/consultant
   2 Independent contractor
   3 Employed – managerial/professional
   4 Employed – clerical

Profile of Household (All)

4 Household type
   1 Solo
   2 Non-related adults
   3 Couple
   4 One-parent
   5 Two-parent
   6 Empty nester
   7 Other _____

I would like to know who else besides yourself lives here, and their relationship to you.

5	Relationship	M	F	Age
_____	_____	__	__	____
_____	_____	__	__	____
_____	_____	__	__	____

6 No. of adults   _____

7 No. of children   _____

Part 2: Profile of Work Patterns (All)

Now, I would like to ask you some questions about your work.

8 What do you do for a living? _____

9 How many people work in your company or government division? _____
   1 Self-employed
   2 Small company
   3 Medium-sized company
   4 Large company
   5 Government division/department
   6 Other _____

10 What do the other adults in your household do?

I want to now ask some questions about your work time schedule.

11 On average, how many days per week do you work?
    Total no. of days   _____

12 On average, how many weeks per year do you work?
    Total no. of weeks _____

13 On average, how many hours per week do you work?
    Total no. of hours _____

(For homeworkers)

14 On average, how many hours per week do you work at home?
    Total no. of hours _____

(All)

15 On average, how many hours per week do you engage in housework?
    Total no. of hours _____

16 I would now like to walk you through a typical working day. When and where are you:
    (a) working, (b) engaged in housework or child care, (c) engaged in leisure or exercise,
    (d) travelling, (e) sleeping, (f) eating? Do you do more than one activity (such as child
    care and work) at once? If so, when do you overlap activities? When are other people
    at home? (OPAH = other people at home)

Time	Activity	Location	OPAH
1 12:00 midnight			
2 1:00 a.m.			
3 2:00 a.m.			
4 3:00 a.m.			
5 4:00 a.m.			
6 5:00 a.m.			
7 6:00 a.m.			
8 7:00 a.m.			
9 8:00 a.m.			
10 9:00 a.m.			
11 10:00 a.m.			
12 11:00 a.m.			
13 12:00 noon			
14 1:00 p.m.			
15 2:00 p.m.			
16 3:00 p.m.			
17 4:00 p.m.			
18 5:00 p.m.			

19 6:00 p.m.
20 7:00 p.m.
21 8:00 p.m.
22 9:00 p.m.
23 10:00 p.m.
24 11:00 p.m.

(For part-time homeworkers)

17 Does your schedule change when you work away from home?
1 Yes
2 No
3 N/a

If so, let's walk through a typical day when you work away from home. When and where are you: (a) working, (b) engaged in housework or child care, (c) engaged in leisure or exercise, (d) travelling, (e) sleeping, (f) eating? Do you do more than one activity (such as child care and work) at once? If so, when do you overlap activities? When are other people at home? (AH = at home, OPAH = other people at home)

Time	Activity	Location	AH	OPAH
1 12:00 midnight				
2 1:00 a.m.				
3 2:00 a.m.				
4 3:00 a.m.				
5 4:00 a.m.				
6 5:00 a.m.				
7 6:00 a.m.				
8 7:00 a.m.				
9 8:00 a.m.				
10 9:00 a.m.				
11 10:00 a.m.				
12 11:00 a.m.				
13 12:00 noon				
14 1:00 p.m.				
15 2:00 p.m.				
16 3:00 p.m.				
17 4:00 p.m.				
18 5:00 p.m.				
19 6:00 p.m.				
20 7:00 p.m.				
21 8:00 p.m.				
22 9:00 p.m.				
23 10:00 p.m.				
24 11:00 p.m.				

18 What do you especially like about your work?

19 Is there anything that you don't like about your work?
1 Yes
2 No
If so, what?

(For homeworkers)

I now want to ask you some questions about working at home.

20 How long have you worked at home?

21 Why did you choose to work at home?

22   Does working at home make your work easier to do?
1   Yes
2   No
If yes, how?

23   Are there any ways which working at home hinders your work?
1   Yes
2   No
3   N/a
If so, how?

24   Does working at home make you *feel* that you have more, less, or about the same amount of control over your work life (i.e., is your work time more or less flexible, do you have more or less autonomy in how you work)?

More   About same   Less   N/a
1   2   3   4

How so?

25   How do you think that the other members of your household feel about you working at home?

26   On a scale from 1 to 5, with 1 being very enjoyable, how much do you enjoy working at home?

Very enjoyable 1   2   3   4   5   Don't enjoy

Part 3: Role of Workplace

Now, I would like to ask you about where you work.

(For homeworkers)

27   Where, in your home, do you do most of your work?
Please describe it.

28   Why did you choose this area to work in?

29   What sets it off as a workplace to others?

30   What makes it feel like a workplace to you?

(For part-time homeworkers and office workers)

31   What kind of office do you work in?
1   High-rise commercial/office
2   Low-rise commercial/office
3   Low-rise mixed use
4   Other _____

32   How is your office laid out?
1   Open plan
2   Ring of offices with steno pool in middle
3   Double-loaded corridor
4   Other _____

(Homeworkers)

33   Does the place where you work in your home make it easy for you to work?
1   Yes
2   No
3   N/a
If so, how does it?

(Office workers)

34 Does the place where you work in your office make it easy for you to work?
1   Yes
2   No
3   N/a
If so, how does it?

(Homeworkers)

35 Is there anything about the place where you work in your home that hinders what you are trying to do in your work?
1   Yes
2   No
3   N/a
If so, what?

(Office workers)

36 Is there anything about the place where you work in your office that hinders what you are trying to do in your work?
1   Yes
2   No
3   N/a
If so, what?

(For homeworkers)

37 Do you use other areas of your home to do your work?
1   Yes
2   No
3   N/a
Is so, what areas do you use?

38 Are there any problems with storing the materials that you need for your work?
1   Yes
2   No
3   N/a
Is so, what are the problems?

39 Have you given any thought to what an ideal home workspace would be like?
1   Yes
2   No
3   Don't know
What would be an ideal home workspace?

40 If you could have your ideal home workplace, would it be very important, somewhat important, or not important to have:

		Very important	Somewhat	Not important
1	Separate entrance to outside	1	2	3
2	Separate room for workspace	1	2	3
3	Workspace in separate structure	1	2	3
4	View out from workspace	1	2	3
5	Workspace that can't be seen from other rooms	1	2	3
6	Private outdoor space attached to workspace	1	2	3

41 If you could have your ideal location for a home workspace what rooms would you like it to be close to?

What rooms would you like it to be far away from?

Part 4: Role of Home and Home Life

(All)

I am now going to ask you some questions about features of your home and home life.

42   Housing type (Researcher fill in)
1   Single detached house
2   Duplex
3   Rowhouse
4   Low-rise apartment (four floors or fewer)
5   High-rise apartment (five floors or more)
6   Mixed use (residential/commercial)
7   Other _____

43   How long have you lived in your present home?

44   Do you own or rent your home?
1   Owner
2   Tenant

45   How many bedrooms do you have in your home?
1   One
2   Two
3   Three
4   Four
5   More than four _____

46   What do you like about your home?

47   Is there anything that you don't like about your home?
1   Yes
2   No
If so, what?

48   I would now like to know how you feel about your home. For example, some people feel that their home is an expression of themselves or is a place where they feel they belong, or they have other different feelings about their home. What feelings do you have about your home?

49   Do you feel relaxed at home?
1   Almost all the time
2   Sometimes
3   Almost never
4   Other _____

50   If it is relaxing, what about your home and home life makes it so?

(For homeworkers)

51   Do you feel different about your home when you work in it?
1   Yes
2   No
3   N/a
If so, in what way?

(All)

52   Do you ever experience conflicts between work and home life?
1   Yes
2   No
If so, what are they?

(For homeworkers)

53 Generally, would you say that working at home has changed the quality of your home life?
   1 Definitely has improved it
   2 Somewhat improved it
   3 About the same as before
   4 Somewhat made it worse
   5 Definitely has made it worse
   6 N/a
   How so?

54 If you could do improvements to your home to make it a better place to work what would they be?

Part 5: Neighbourhood Use

(All)

Now, I would like to ask you some questions about the use of your neighbourhood.

55 Neighbourhood type (researcher fill in)
   1 Urban
   2 Mature suburban
   3 Suburban
   4 Other _____

56 How would you describe your neighbourhood, as a place to live, to a friend?

57 What would you say are the boundaries of the neighbourhood that you live in?

58 What kinds of people live in your neighbourhood?

59 What things do you typically do in your neighbourhood?

60 Is your neighbourhood changing in any way?
   1 Yes
   2 No
   If so, how is it changing?

61 Are you friendly with people in your neighbourhood?
   1 Yes
   2 No

   If so, how often do you see them?
   1 Frequently (at least once a week)
   2 Occasionally (at least once a month)
   3 Rarely
   4 Other _____

62 In general, how often do you see your friends who live outside of your neighbourhood?
   1 Frequently (at least once a week)
   2 Occasionally (at least once a month)
   3 Rarely
   4 Other _____

63 Where do you usually see your friends?
   1 My house
   2 At restaurant
   3 At friend's house
   4 Other _____

(All)

64 Do you use the following services a lot (on a daily to weekly basis), sometimes (on a weekly to monthly basis), or rarely (on a monthly or less basis)?

		A lot	Sometimes	Rarely	N/a
1	Shops	1	2	3	4
2	Parks	1	2	3	4
3	Children's playgrounds	1	2	3	4
4	Daycare centre	1	2	3	4
5	Copy centre	1	2	3	4
6	Post office	1	2	3	4
7	Cafés/restaurants	1	2	3	4
8	Recreation centre/gym	1	2	3	4
9	Other _____	1	2	3	4

(For homeworkers)

65 When you work at home do you use the following neighbourhood services more, less, or about the same as you do when you are not working at home?

		More	About same	Less	N/a
1	Shops	1	2	3	4
2	Parks	1	2	3	4
3	Children's playgrounds	1	2	3	4
4	Daycare centre	1	2	3	4
5	Copy centre	1	2	3	4
6	Post office	1	2	3	4
7	Cafés/restaurants	1	2	3	4
8	Recreation centre/gym	1	2	3	4
9	Other _____	1	2	3	4

66 When you work at home do you notice what is going on in your neighbourhood, such as noticing strangers on your street or children playing, more, less, or about the same as you do when you are not working at home or before you worked at home?

More	About same	Less	N/a
1	2	3	4

67 When you work at home do you socialize with your neighbours more, less, or about the same as you do when you are not working at home or before you worked at home?

More	About same	Less	N/a
1	2	3	4

68 On the days when you work at home do you spend more, less, or about the same amount of your leisure time in your neighbourhood than on those days when you work elsewhere or before you worked at home?

More	About same	Less	N/a
1	2	3	4

69 Do you feel different about your neighbourhood on the days when you work at home than on the days that you work elsewhere or before you worked at home?
1 Yes
2 No
3 N/a
If so, in what way do you feel different about your neighbourhood?

70 Are there any improvements that you would like to see in your neighbourhood to make it better for you to work at home?
1 Yes
2 No
3 N/a
If so, what are the improvements?

Part 6: Assessment of Working at Home

(All)

71 Generally, what do you think are the benefits of working at home?

72 What do you think are the disadvantages of working at home?

(For homeworkers)

73 Would you recommend working at home to other people?
1 Yes
2 No
3 N/a

74 Do you expect your working at home arrangement to continue for the foreseeable future?
1 Yes
2 No
3 N/a
If not, where do you see yourself working?

(For office workers)

75 Would you like to work at home?
1 Yes
2 No
3 N/a
If so, why?
If not, why not?

(All)

76 If you would like to work at home would you like it to be full-time, part-time, flextime, or some other arrangement?
1 Full-time
2 Part-time
3 Flextime
4 Other _____
Why would you prefer that arrangement?

(All)
77 Would you be interested in working in a satellite office close to your home?
1 Yes
2 No
If so, why?

(For homeworkers)

78 What advice would you give to someone planning to work at home?

Part 7: Telecommunications Usage

(All)

I now want to ask you some questions about your use of office equipment.

79 What kind of office equipment do you use on a daily basis?
1 Telephone
2 Computer terminal
3 Modem
4 Printer
5 Facsimile machine
6 Photocopier

7   Telephone answering machine
8   Voice mail
9   Electronic mail
10 Postal meter
11 Other _____

(For homeworkers)

80 What other equipment would you like to have in your home office?

(All)

81 Do you prefer to work with people face to face, by telephone, or some other way?

		Face to face	Telephone	Other way
1	Clients	1	2	3
2	Employers	1	2	3
3	Employees	1	2	3
4	Business associates	1	2	3

82 Are there advantages to discussing work over the telephone or sending information via a computer over face-to-face contacts?
1   Yes
2   No
3   Don't know
If so, what are they?

83 Are there disadvantages to discussing work over the telephone or sending information via a computer over face-to-face contacts?
1   Yes
2   No
3   Don't know
If so, what are they?

Part 8: Demographics of Respondent

This final set of questions concerns some general aspects about yourself, and I would like you to fill them in yourself.

84 What is the highest level of education you have completed?
1   Some high school
2   Completed high school
3   Technical school
4   Some college
5   Completed college
6   Attended graduate school
7   Completed graduate school
8   Other _____

85 What was your approximate *personal* income in 1988 from all sources before taxes?
1   Less than $10,000
2   $10,000 to $14,999
3   $15,000 to $19,999
4   $20,000 to $29,999
5   $30,000 to $39,999
6   $40,000 to $49,999
7   $50,000 to $74,999
8   $75,000 to $99,999
9   $100,000 and over

86 What was your approximate *household* income in 1988 from all sources before taxes?
   1 Less than $10,000
   2 $10,000 to $14,999
   3 $15,000 to $19,999
   4 $20,000 to $29,999
   5 $30,000 to $39,999
   6 $40,000 to $49,999
   7 $50,000 to $74,999
   8 $75,000 to $99,999
   9 $100,000 and over

87 What is your age?
   1 20-9
   2 30-9
   3 40-9
   4 50-9
   5 60-9
   6 70 and over

(For homeworkers)

I would like to now sketch your workplace and take some photographs.

Thank you for your time.

Other comments:

Time of interview:

Location in home:

Photos taken: front façade, character of street, home office interior

Sketches done by interviewer:

Features of Neighbourhood, Home, and Workspace
(To be completed by researcher)

1 Location of neighbourhood
   1 In a large city (250,000+)
   2 In a medium-sized city (50,000-249,000)
   3 In a small city or town (under 50,000)
   4 In a suburb near a large city
   5 In a suburb near a medium-sized city
   6 In a suburb near a small city
   7 Other _____

2 Street character of respondent's neighbourhood
   1 Residential
   2 Commercial
   3 Mixed
   4 Other _____

3 If more than one dwelling unit, is the access from the street
   1 Shared entrance
   2 Private entrance

4 Is there a separate entrance to the office from outside?
   1 Yes
   2 No

5   Access to workspace can be closed
    1   Yes
    2   No

6   Presence of personal, non-work items (plants, art, coffee pot)

7   Outside views from workspace room.

Sketch of workspace to include
    1   Room layout: approximate dimensions and north orientation
    2   Doors and windows
    3   Furniture arrangement (table, desk, file cabinet, bookshelves, chair, sofa, other furniture)
    4   Equipment arrangement (computer, telephone, answering machine, typewriter, radio, clock, television, other)
    5   Sources of artificial light
    6   Adjacent rooms (those with visual and physical access)

Appendix B
Canadian Telework and Home-Based Employment Survey

Some Special Use of Words

The survey uses some words in a special way:

Teleworker/homeworker/telecommuter is someone who works from home, part or full time, as an employee for a public institution, Crown corporation, or private corporation.

Independent contractor is someone who works from home, part or full time, on contract to one company such as a contract employee or piece worker.

Self-employed consultant is someone who works from home, part or full time, doing consulting work for more than one company or individual.

Home-based business operator is someone who works from home, part or full time, providing a service or product to a variety of clients or customers.

Moonlighter is someone who works from home on a part-time basis as a supplemental job in addition to a primary job.

Occasional homeworker is someone who brings work home after work hours from the workplace on a frequent to occasional basis.

Full-time home-based worker works all of his or her time at home.

Part-time home-based worker works four days or less at home and the rest of the time at another work location.

Neighbourhood telework centre is office space shared by a number of unrelated businesses located in a convenient location in a neighbourhood .

Satellite office is a company's secondary office located close to employees' homes; intended to reduce the commute to the corporate head office.

General

To begin, please answer some general questions about yourself as a home-based worker so that we can understand the range of home-based workers across the country.

1 Which of the following statements best describes your current situation? (Check the one that best applies to your situation.)

☐ I work at home as a *teleworker/homeworker/telecommuter* (i.e., someone who works from home, part or full time, as an employee for a public institution or Crown corporation).

☐ I work at home as a *teleworker/homeworker/telecommuter* (i.e., someone who works from home, part or full-time, as an employee for a private corporation).

☐ I work at home as an *independent contractor* (i.e., someone who works from home, part or full time, on contract to one company such as a contract employee or piece worker).

☐ I work at home as a *self-employed consultant* (i.e., someone who works from home, part or full time, doing consulting work for more than one company or individual).

☐ I work at home as a *home-based business operator* (i.e., someone who works from home, part or full time, providing a service or product to a variety of clients or customers).

☐ I work at home as a *moonlighter* (i.e., someone who works from their home on a part-time basis as a supplemental job in addition to a primary job).

☐ I work at home as an *occasional homeworker* (i.e., someone who brings work home after work hours from the workplace on a frequent to occasional basis).

☐ Other

☐ I do not work at home.

If you are not working at home please stop here and send the survey back to us in the enclosed stamped envelope. Thank you for your time.

If you are working at home please continue filling out the survey.

---

2 On average, how many hours per week do you do:
   work for income at home?
   work for income at another work location? _____

3 On average, how many hours per week do you do:
   household chores/maintenance?   _____
   child care?   _____
   leisure activities/volunteer work?   _____

4 In the past week, what percentage of your work time did you spend at the following locations? (Please write your best estimate in each space provided.)
   _____ % in home office or workshop
   _____ % travelling (i.e., travelling to job sites, travelling out of town)
   _____ % in every other work-related place (i.e., at client's office or home, at suppliers, at alternative work site, at seminars, trade shows, etc.)

5 If you are a teleworker/homeworker/telecommuter are you in:
   ☐ a formal telework program? Duration of program/participation:
      _____ years   _____ months
   ☐ an informal arrangement? Duration:
      _____ years   _____ months

6 Do you need child care while you are working at home?
   ☐ Yes   ☐ No

   If so, what type of child care do you use?
   ☐ Child care centre
   ☐ Babysitter
   ☐ Nanny
   ☐ A relative
   ☐ Spouse does it
   ☐ I do it
   ☐ Other (please specify): _____

   Comments: _____
   _____
   _____
   _____

Work Activities

The following questions are about the work you do, so that we can understand the range of work being conducted at home.

7 What is (are) your current home-based occupation(s)/area(s) of employment?

_____

8 If your home-based occupation is not your primary employment, please list your primary employment (i.e., other than your home-based work.)

_____

9 If you had another occupation/area of employment before your home-based work, what was it?

_____

10 How would you categorize your home-based occupation/area of employment? (For those who work at home as a supplemental job, i.e., in addition to another job, please answer the following questions pertaining to your supplemental job.)
   ☐ Agriculture/primary industry
   ☐ Construction and trades
   ☐ Manufacturing/processing (food, textile, arts, crafts, etc.)
   ☐ Transportation/public utilities
   ☐ Wholesale trade
   ☐ Retail trade/product sales
   ☐ Finance, insurance, real estate
   ☐ Personal services (cleaning, hairdressing, etc.)
   ☐ Business services (computing, word processing, consulting, design services, etc.)
   ☐ Health, social, recreational services
   ☐ Administrative services (policy analyst, etc.)
   ☐ Professional services (accounting, educational, legal, architecture, etc.)
   ☐ Other

11 How many months/years have you been:
   in this occupation/area of employment?   ____ years ____ months
   working at home in this occupation/ area of employment?   ____ years ____ months

12 Is your home-based work of a seasonal nature (i.e., you work at it only a few months of the year)?
   ☐ Yes   ☐ No
   If yes, how many months per year do you do home-based work? _____

13 Why did you start working at home? (Check all that apply.)
   ☐ Convenience (e.g., lack of commute)
   ☐ Cost (e.g., reduced overhead)
   ☐ Change in family situation
   ☐ Change in work status (e.g., loss of job; change from employee to contract employee)
   ☐ Health problems
   ☐ Wanted flexible hours
   ☐ Wanted control over work
   ☐ Wanted control over work environment (e.g., need for small workspace)
   ☐ Always worked at home
   ☐ Other (please specify): _____

14 Do you employ others in your home workspace?
   ☐ Yes   ☐ No

   If so, how many people do you employ?
   _____ part-time
   _____ full-time

Home Environment

The following questions are about your home environment, so that we can understand how homes are used by home-based workers and how they can be improved to make them more suitable for home-based work.

15 What type of home do you live in?
   ☐ Single detached house
   ☐ Duplex/semi-detached
   ☐ Rowhouse/townhouse
   ☐ Low-rise apartment (four floors or fewer)
   ☐ High-rise apartment (five floors or more)
   ☐ Mixed-use building (i.e., residential/commercial)
   ☐ Trailer
   ☐ Other (please specify): _____

16 What is your housing tenure?
   ☐ Own home/freehold
   ☐ Condominium/strata title
   ☐ Rent
   ☐ Co-operative
   ☐ Room and board
   ☐ Other _____

17 What is the age of your home? _____ years

18 How long have you lived in your present home?
   _____ years   _____ months

19 How many bedrooms do you have in your home? _____ bedrooms

20 What is the approximate floor area of your home? _____ m² or _____ square feet

21 What was last month's rent or mortgage payment for your home? $ _____

22 For which of the following work-related activities do you use your home?
   (Choose as many as apply.)
   ☐ To produce/manufacture goods
   ☐ To produce/provide services
   ☐ To make and receive business
      telephone calls
   ☐ As a mailing address
   ☐ To do administrative work
   ☐ To store goods/products
   ☐ To store equipment
   ☐ For client/customer meetings
   ☐ Business use of outside property
      (yard, driveway, etc.)
   ☐ To produce/provide professional
      services
   ☐ Other_____

23 Do you have a designated office and/or workshop in your home?
   ☐ Yes   ☐ No
   If yes, where is it located? _____
   What is the approximate floor area? _____m² or _____ sq. ft.

24 How often do you use each of the following rooms in your home for your work?
   (Please indicate the extent to which the following rooms are used for your work by
   telling us whether they are used: never, seldom, frequently, or always. For each room,
   please circle the number that best describes the frequency of use. If you do not have a
   particular room, please circle the "Don't have" category.)

	Frequency of use for work				
	Never	*Seldom*	*Frequently*	*Always*	*Don't have*
Kitchen/dining room	1	2	3	4	5
Living room	1	2	3	4	5
Bedroom	1	2	3	4	5
Rec/family room	1	2	3	4	5

Office/study	1	2	3	4	5
Workshop	1	2	3	4	5
Basement	1	2	3	4	5
Enclosed porch/balcony	1	2	3	4	5
Attached garage	1	2	3	4	5
Detached building	1	2	3	4	5
Other	1	2	3	4	5

25 Are the rooms that you use for work-related activities still used for their original in-tended purpose? (Please list your three most frequently used rooms for work-related functions and circle whether they are used for other purposes: never, seldom, fre-quently, or always.)

☐ Room has always been used for work-related activities (go to question 26).

	Used for original intended purpose			
Room used for work	*Never*	*Seldom*	*Frequently*	*Always*
_____	1	2	3	4
_____	1	2	3	4
_____	1	2	3	4

26 Why did you choose the space in your home that you primarily work in? (Please check all that apply.)
☐ It was available (i.e., only room available; not being used)
☐ It was easily convertible for work activities
☐ In the right location to minimize impact on rest of home
☐ View to outside/natural lighting
☐ Close to entry
☐ It was already equipped with phone jacks and electrical outlets
☐ Don't know
☐ Other (please describe): _____

27 What physical changes have you done or would like to do to your home to make it more suitable for work-related activities? (Please check those changes you have done or would like to do to your home.)

	Done	*Would like to do*	*N/a*
New carpeting	☐	☐	☐
External painting/siding	☐	☐	☐
Internal painting/wall covering	☐	☐	☐
New lighting	☐	☐	☐
New electrical circuitry	☐	☐	☐
New plumbing	☐	☐	☐
New heating system	☐	☐	☐
New ventilation/air conditioning	☐	☐	☐
Erection of walls/doors	☐	☐	☐
Removal of walls/doors	☐	☐	☐
Addition of new room(s)	☐	☐	☐
Renovation/finishing of room(s)	☐	☐	☐
Installation of ramp	☐	☐	☐
Addition of detached/attached building	☐	☐	☐
Landscaping	☐	☐	☐
Other (please specify): _____	☐	☐	☐

28 Do you have any disabilities?
☐ Yes ☐ No
If yes, what work-related modifications have you made to your home to accommodate your disability? _____

29 Have you had any of the following problems when working at home? (Please specify whether each of the following is no problem, a moderate problem or a serious problem for you.)

	No problem	Moderate problem	Serious problem	Don't know	N/a
Noise disturbances from outside your workspace	☐	☐	☐	☐	☐
Inadequate employee/visitor/client parking	☐	☐	☐	☐	☐
Lack of storage for materials/products	☐	☐	☐	☐	☐
Poor electrical wiring	☐	☐	☐	☐	☐
Poor lighting	☐	☐	☐	☐	☐
Inadequate number of telephone lines	☐	☐	☐	☐	☐
Poor ventilation	☐	☐	☐	☐	☐
Lack of space for loading/unloading/ delivery of materials/finished products	☐	☐	☐	☐	☐
Work-related materials stored in home are hazardous	☐	☐	☐	☐	☐
Work-related activities incompatible with home environment	☐	☐	☐	☐	☐
Workspace too small	☐	☐	☐	☐	☐
Intrusions from family/neighbours/ friends	☐	☐	☐	☐	☐
Opposition from neighbours to work activities	☐	☐	☐	☐	☐
Complaints from municipal agencies re zoning infractions, incompatible uses, etc.	☐	☐	☐	☐	☐
Other_____	☐	☐	☐	☐	☐

How have you addressed some of your major problems working at home?

_____

_____

30 If you could have your ideal home workspace would it be very important, somewhat important, or not important to have the following features in it:

	Very important	Somewhat important	Not important	Don't know
A separate entry from the street	☐	☐	☐	☐
Separate room for workspace	☐	☐	☐	☐
Workspace in separate structure	☐	☐	☐	☐
Natural lighting (i.e., window)	☐	☐	☐	☐
Visual privacy from other activities	☐	☐	☐	☐
Acoustical privacy from other activities	☐	☐	☐	☐
Availability of exit doors and phones in case of emergency	☐	☐	☐	☐
Adequate storage space	☐	☐	☐	☐
Adequate electrical amperage and number of outlets	☐	☐	☐	☐
Proper ventilation	☐	☐	☐	☐
Large doors for ease of loading/ unloading of materials	☐	☐	☐	☐
Load-bearing floors	☐	☐	☐	☐
Storage for hazardous materials	☐	☐	☐	☐
Other _____	☐	☐	☐	☐

31 Since you have started working at home do you feel your household has generated more, less, or about the same amount of the following: (Please circle the appropriate amount next to each item that reflects your assessment.)

	More	Less	Same amount	Don't know	N/a
Noise	☐	☐	☐	☐	☐
Sewage	☐	☐	☐	☐	☐
Garbage	☐	☐	☐	☐	☐
Traffic	☐	☐	☐	☐	☐
Odours	☐	☐	☐	☐	☐
Chemical waste	☐	☐	☐	☐	☐

32 Since working at home, have you ever thought of moving or actually moved because: (Check as many as apply.)

	Have moved	Thought of moving	N/a
Need more space for work activities	☐	☐	☐
Need more separation between home and work activities	☐	☐	☐
Home not matching business image	☐	☐	☐
Lack of privacy	☐	☐	☐
Lack of safety/security	☐	☐	☐
Complaints re: work activities	☐	☐	☐
Working at home gave you the freedom to live where you want	☐	☐	☐
Other _____	☐	☐	☐

Comments: _____

_____

33 If you did move or you want to move, what kind of home did you move to or do you want to move to, and where is it or where would it be located? (Please check as many as apply for type of home and location of home.)

Type of home
☐ Larger home
☐ Smaller home
☐ More appropriate layout
☐ More amenities
☐ Other _____

Location of home
☐ Same neighbourhood
☐ Same city
☐ Closer to city centre
☐ Further from city centre
☐ Outside city
☐ Other _____

Comments: _____

_____

34 Please indicate whether you currently use the following equipment or services to conduct your home-based work activities. (Check as many as apply.)
☐ Business tel. no.
☐ Residential tel. no.
☐ Computer
☐ Printer
☐ Fax machine
☐ Cellular phone
☐ Typewriter
☐ Pager
☐ Photocopier
☐ Telephone answering machine/voice mail
☐ Electronic mail/networks
☐ Other

35 Other than the above, what other pieces of equipment or large tools do you have in your home workspace?

_____

_____

36 What equipment/services that you don't have would you like to have in your home workspace?

_____

_____

Community Environment

The following questions are about the community that you live in. We are asking these questions so that we can understand how home-based workers use their communities and how they can be improved to make them more convenient for home-based work.

37   How would you describe the type of community that you live in?
☐ Urban (i.e., central city neighbourhood, multiple family housing)
☐ Urban (i.e., single family housing)
☐ Mature suburban (i.e., outside urban core and ten years or older)
☐ New suburban (i.e., outside urban core and less than ten years old)
☐ Small town (i.e., under 20,000 population)
☐ Rural area
☐ Other _____

38   What is the character of your street?
☐ Residential
☐ Commercial
☐ Industrial
☐ Mixed use
☐ Rural
☐ Other _____

39   What of the following services are available in your community within a ten-minute walk or a five-minute drive from your home and when you work at home do you use them more, less, or about the same as when you are not working at home or before you worked at home? (Check whether the service is available in your community and whether you use it more, less, or the same as when not working at home.)

	Available	If available, use:			
		More	Less	Same	N/a
Shops/personal services	☐	☐	☐	☐	☐
Parks	☐	☐	☐	☐	☐
Banks	☐	☐	☐	☐	☐
Copy centre	☐	☐	☐	☐	☐
Post office	☐	☐	☐	☐	☐
Cafés/restaurants	☐	☐	☐	☐	☐
Recreation centre/gym	☐	☐	☐	☐	☐
Childcare centre	☐	☐	☐	☐	☐
Other _____	☐	☐	☐	☐	☐

40   When you work at home do you notice what is happening on your street (such as noticing strangers on your street or children playing) more, less, or about the same as when you are not working at home or before you worked at home?

| More | Same | Less | N/a |
| 1 | 2 | 3 | 4 |

41   When you work at home do you interact with your neighbours more, less, or about the same as when you are not working at home or before you worked at home?

| More | Same | Less | N/a |
| 1 | 2 | 3 | 4 |

42   When you work at home do you use the following modes of transportation for work and other activities more, less, or about the same as when you are not working at home or before you worked at home.

	More	Same	Less	N/a
Automobile	1	2	3	4
Public transportation	1	2	3	4
Bicycle	1	2	3	4
Walk	1	2	3	4
Other _____	1	2	3	4

43 In your last working week, what was the average distance per day (in km or mi.) you travelled:

When you were not working at home		*When you worked at home*	
For work activities	_____ km or _____ mi.	_____ km or _____ mi.
For household chores, etc.	_____ km or _____ mi.	_____ km or _____ mi.
For leisure activities	_____ km or _____ mi.	_____ km or _____ mi.
All activities	_____ km or _____ mi.	_____ km or _____ mi.

44 On a scale of 1 to 5, with 1 being very negative and 5 being very positive, how do you feel that having work-related activities in your home has affected the residential atmosphere of your neighbourhood?

Very negative		*No effect*		*Very positive*	*N/a*	*Don't know*
1	2	3	4	5	☐	☐

Comments: _____

45 What municipal regulations have you dealt with as a result of your home-based work activities? (Please check those regulations or boards that you have encountered that pertain to your home-based work activities.)
☐ Building permit
☐ Inspectors (building, health, safety)
☐ Business licence
☐ Sign by-law
☐ Zoning variance
☐ Parking requirements
☐ Fencing, landscaping requirements
☐ Public meeting for approval of work activity
☐ Public meeting to deal with complaints about your work activity
☐ Other (please specify): _____

Comments: _____

46 Are there any improvements such as more services, changes to zoning, etc., that you would like to see in your neighbourhood to make it more convenient for you to work at home?
☐ Yes   ☐ No   ☐ N/a   ☐ Don't know

If so, what improvements would you like to see? _____

Home Life

The following general questions are about how you feel about working at home and its effects on your home life.

47 Generally, what do you think are the benefits to working at home?

_____

_____

48 What do you think are the disadvantages to working at home?

_____

_____

49 Please look at this list of statements some home-based workers have made about the relationship between their work and their home life. For each one indicate on a scale of 1 to 5 whether you agree or disagree that it applies to you and your household. (Circle one number for each statement. If the statement does not apply to your household, circle N/a.)

	Strongly disagree				Strongly agree	
I have more time to devote to household responsibilities and personal business than I would if I worked from a location away from home.	1	2	3	4	5	N/a
I get distracted from work activities by household responsibilities and chores.	1	2	3	4	5	N/a
My work life would be less stressful if I worked from a location away from home.	1	2	3	4	5	N/a
I often feel lonely when working at home.	1	2	3	4	5	N/a
I have fewer friends since working at home.	1	2	3	4	5	N/a
I miss the camaraderie of a workplace.	1	2	3	4	5	N/a
I work too much when I work at home.	1	2	3	4	5	N/a
Other members of the household and/or friends frequently interrupt my work activities.	1	2	3	4	5	N/a
My time with members of my household is often interrupted by work activities.	1	2	3	4	5	N/a
Other members of the household and/or friends have learned a great deal about my work.	1	2	3	4	5	N/a
The lives of household members would be less stressful if I worked from a location away from home.	1	2	3	4	5	N/a
I know my neighbours better when I work at home.	1	2	3	4	5	N/a
I spend my leisure time close to home when I work at home.	1	2	3	4	5	N/a

Comments:_____

_____

50 On a scale of 1 to 5, with 1 being very dissatisfied and 5 being very satisfied, how satisfied are you with working at home?

Very dissatisfied		*Somewhat satisfied*		*Very satisfied*	*Don't know*
1	2	3	4	5	☐

Comments:_____

_____

51 Do you foresee working at home as a long-term employment strategy for you?
☐ Yes   ☐ No   ☐ Don't know

If so, why or why not?_____

52 Would you be interested in working from a neighbourhood telework centre or satellite office site? (A *neighbourhood telework centre* is office space shared by a number of unrelated businesses in a convenient location in a neighbourhood. A *satellite office* is a company's secondary office located close to employees' homes, intended to reduce the commute to the corporate head office.)
☐ Yes   ☐ No   ☐ N/a   ☐ Don't know

If so, why or why not? _____

Demographic Information

To conclude, we would like to ask you some additional questions about yourself and your household so that we can understand the range of households with home-based work activities. All your answers will be confidential and your name will not appear in any report.

53 Please tell us the age and sex of each person who normally lives in your home, and their relationship to you:

	Age	Sex Male	Female	Relation to you Husband/wife /common-law	Child	Other relation	Other
Yourself	_____	☐	☐	☐	☐	☐	☐
Person 2	_____	☐	☐	☐	☐	☐	☐
Person 3	_____	☐	☐	☐	☐	☐	☐
Person 4	_____	☐	☐	☐	☐	☐	☐
Person 5	_____	☐	☐	☐	☐	☐	☐
Person 6	_____	☐	☐	☐	☐	☐	☐
Other	_____	☐	☐	☐	☐	☐	☐

54 What is the highest level of education you have completed?
☐ Grade school
☐ High school
☐ Some post-secondary school
☐ Technical school
☐ College/university
☐ Graduate school
☐ Other _____

55 How many years have you been in the paid workforce? _____

56 What category does your total household income fall under?
☐ Under $15,000
☐ $15,000 to $29,000
☐ $29,001 to $58,000
☐ $58,001 to $85,000
☐ Over $85,001

57 What percentage of your total household income is from your home-based work?

This concludes our questionnaire. Thank you for your time.

Appendix C
Respondent Occupations, California Study

Full-time Homeworkers

Male
1 Government manager
2 Architect
3 Writer
4 Radio producer/writer
5 Technical writer/video producer
6 Journalist
7 Dealer in out-of-print books
8 Editor/publisher
9 Computer programmer
10 Documentary filmmaker
11 Electrical engineering consultant

Female
1 Technical writer
2 Organizational consultant
3 Documentary filmmaker
4 Medical transcriptionist
5 Technical writer/market researcher
6 Medical transcriptionist
7 Word processor
8 Interior designer
9 Writer/editor
10 Writer
11 Word processor
12 Data-processing consultant
13 Editor
14 Writer/public relations

Part-time Homeworkers

Male
1 Government management analyst
2 Government community social service analyst
3 Technical writer/editor
4 Government program developer
5 Government juvenile justice consultant
6 Business consultant
7 Architect

Female
1 Health planner
2 Government administrator/ management analyst
3 Electrical engineer
4 Medical transcriptionist
5 Computer analyst
6 Government community services consultant
7 Government stenographer
8 Government research analyst

Office Workers

Male
1 University senior planner
2 Government financial and regulatory analyst
3 Government senior environmental planner
4 Government staff programming analyst

Female
1 Government information systems analyst
2 Government planner
3 Government attorney
4 Technical writer/editor
5 Medical transcriptionist

Appendix D
Respondent Occupations, Canadian Survey

Occupation of Home-Based Work	#	%
1 Agriculture	**2**	**0.33**
1.1 Farming	1	0.17
1.2 Nursery/greenhouse	1	0.17
2 Construction and trades	**5**	**0.83**
2.1 Construction	1	0.17
2.2 Trades	4	0.66
3 Manufacturing and processing	**71**	**11.77**
3.1 Arts/crafts	48	7.96
3.2 Food	4	0.66
3.3 Software producer	6	1.00
3.4 Clothing/dressmaking	5	0.83
3.5 Other	8	1.33
4 Transportation/public utilities	**0**	**0.00**
5 Wholesale/trade	**5**	**0.83**
6 Retail trade and product sales	**98**	**16.25**
6.1 Retail	54	8.95
6.2 Product sales/sales representative	44	7.30
7 Finance, insurance, real estate	**27**	**4.48**
7.1 Finance	3	0.50
7.2 Insurance agent	2	0.33
7.3 Insurance assessing	17	2.82
7.4 Real estate development	4	0.66
7.5 Real estate mortgagor	1	0.17
8 Personal services	**20**	**3.32**
8.1 Cleaning	3	0.50
8.2 Hairdressing	5	0.83
8.3 Dog training	1	0.17
8.4 Gardening	3	0.50
8.5 Other	8	1.33

▶

Occupation of Home-Based Work	#	%
9 Business services	**198**	**32.84**
9.1 Computer services/programming	38	6.30
9.2 Word processing	18	2.99
9.3 Graphic design service	16	2.65
9.4 Design service	4	0.66
9.5 Training	15	2.49
9.6 Financial management	6	1.00
9.7 Marketing	22	3.65
9.8 Bookkeeping	10	1.66
9.9 Writing/publishing	42	6.97
9.10 Conference planner	4	0.66
9.11 Video production/photography	8	1.33
9.12 Other	15	2.49
10 Health, social, and recreational	**27**	**4.48**
10.1 Entertainer	3	0.50
10.2 Health, nutrition	5	0.83
10.3 Child care/foster parent	2	0.33
10.4 Resort/lodge operator	6	1.00
10.5 Social service consultant	9	1.49
10.6 Other	2	0.33
11 Administrative	**40**	**6.63**
11.1 Government analyst	13	2.16
11.2 Government researcher	9	1.49
11.3 Government manager	8	1.33
11.4 Other	10	1.66
12 Professional services	**110**	**18.24**
12.1 Chartered accountant	3	0.50
12.2 Education	88	14.59
12.3 Architect/landscape architect/interior designer/planner	4	0.66
12.4 Lawyer	1	0.17
12.5 Engineer	8	1.33
12.6 Environmental consultant	3	0.50
12.7 Other	3	0.50
Total	**603**	**100.00**

Note: Of the total number of respondents (n = 453), forty-seven listed two occupations and twelve listed three to five occupations, making a total of 603 occupations cited. Percentages reflect rounding.

Notes

Chapter 3: Working at Home and Being at Home

1 The following profiles of home-based workers amplify the socioeconomic trends outlined in Chapter 2. Among the patterns of daily life profiled are the use of time and space in a typical day, work patterns, personality attributes, and the habits of daily life such as diet, exercise, and socializing. The profiles are drawn from qualitative and quantitative data derived from the 1990 California sample, which included a comparison sample of home-based workers and office workers.

2 The 1990 study formed the basis for the author's PhD dissertation, "Working at Home in the Live-In Office: Computers, Space, and the Social Life of Households," University of California, Berkeley.

Chapter 4: A Strategy of a Dispensable Workforce

1 The 1995 survey by the author, "Planning for Telework and Home-Based Employment: A Canadian Survey on Integrating Work into Residential Environments," was supported by Canada Mortgage and Housing Corporation (CMHC), a Crown corporation of the Federal Government of Canada.

Chapter 5: Localizing the Networked Economy

1 The data for this chapter were obtained from a case study of the Vancouver region, developed by the author. This project used multiple methods of quantitative and qualitative analysis to investigate the extent of telework in the greater Vancouver area and the experience of telework for selected teleworkers. The methodology incorporated: (1) analysis of employment and journey to work data in the greater Vancouver area from the 1991 and 1996 censuses to determine the work patterns of people in nontraditional workplaces (such as homes); (2) verification and update of data on companies offering telework programs; and (3) key informant interviews with eight selected teleworkers, four of whom were "knowledge workers" and four manual workers (i.e., they did data entry). The interviews were structured to obtain data on how teleworkers see their roles as workers and in the domestic sphere, and their use of space and time. Informal interviews were also conducted with teleworkers found through chance meetings. The questionnaire instrument was an amalgamation of the questionnaires from the 1990 California study and the 1995 Canada-wide survey, updated to the year 2000 context.

References

Adair, John G. 1984. "The Hawthorne Effect: A Reconsideration of the Methodological Artifact." *Journal of Applied Psychology* 69: 334-45.

Ahrentzen, Sherry. 1987. "Blurring Boundaries: Socio-Spatial Consequences of Working at Home." Working paper, Department of Architecture, University of Wisconsin-Milwaukee, Milwaukee.

–. 1989. "A Place of Peace, Prospect, and ... a PC: The Home As Office." *Journal of Architecture and Planning Research* 6 (Winter): 4.

Aldrich, M. 1982. *Videotex: Key to the Wired City.* London: Penguin.

Allen, S., and C. Wolkowitz. 1987. *Homeworking: Myths and Realities.* Basingstoke, UK: Macmillan Education.

"Analysis of Data from National Survey Shows Surprising Trends, Raises Interesting Questions about Home Work." 1988. *Telecommuting Review* (February): 12-17.

Andrew, Caroline. 1992. "The Feminist City." In *Political Arrangements and the City,* edited by Henri Lustiger-Thaler. Montreal: Black Rose Books, 109-22.

Antonoff, M. 1985. "The Push for Telecommuting." *Personal Computing* 9 (7): 82-92.

Armstrong-Stassen, M., M. Solomon, and A. Templay. 1998. "Alternative Work Arrangements: Meeting the Challenges." *Canadian Psychology* 39 (1/2): 108-23.

Aslanbeigui, Nahid, Steven Pressman, and Gale Summerfield, eds. 1994. *Women in the Age of Economic Transformation: Gender Impact of Reforms in Post-Socialist and Developing Countries.* London: Routledge.

Atkinson, J. 1984. *Flexibility, Uncertainty and Manpower Management.* Brighton, UK: Institute of Manpower Studies.

Atkinson, W. 1985. *Working at Home? Is It for You?* Homewood, IL: Down Jones-Irwin.

Bailey, Brian. 1998. *The Luddite Rebellion.* Phoenix Mill, UK: Sutton Publishing.

Baran, Barbara. 1985. "Office Automation and Women's Work: The Technological Transformation of the Insurance Industry." In *High Technology, Space, and Society,* edited by M. Castells, 143-71. Beverly Hills, CA: Sage.

Barbineau, Guy. 2000. "The Full Read on Employment: Brave New Workplace." *Vancouver Sun,* 5 February, A6.

Baxter, David. 1994. "Changes in Journey to Work Patterns in Metropolitan Vancouver: 1971 to 1991." Report. Greater Vancouver Regional District, BC.

Beecher, Catherine E., and Harriet Beecher Stowe. [1869] 1975. *The American Woman's Home.* Hartford, CT: Stowe-Day Foundation.

Bell, Daniel. 1973. *The Coming of Post-Industrial Society.* New York: Basic Books.

Bellamy, Edward. [1888] 1967. *Looking Backward 2000-1887.* Cambridge, MA: Harvard University Press.

Bender, Thomas. 1978. *Community and Social Change in America.* New Brunswick, NJ: Rutgers University Press.

Berger, Marguerite. 1989. "Giving Women Credit: The Strengths and Limitations of Credit As a Tool for Alleviating Poverty." *World Development* 17: 1017-32.

Bibby, Andrew. 1999. "Teleworking: How the Trade Unions Are Responding." <www.eclipse. co.uk/pens/bibby/hw-tu.html>.

Blakeley, Edward J., and Mary Gail Snyder. 1997. *Fortress America: Gated Communities in the United States*. Washington, DC: Brookings Institution Press.

Boei, William. 1999. "Nine Firms Fishing for Wireless Net Licence." *Vancouver Sun*, 12 November, F1.

Boris, Eileen. 1994. *Home to Work: Motherhood and the Politics of Industrial Homework in the United States*. Cambridge: Cambridge University Press.

Boris, Eileen, and C. Daniels. 1989. *Homework: Historical and Contemporary Perspectives on Paid Labor at Home*. Urbana: University of Illinois Press.

Boris, Eileen, and E. Prügl, eds. 1996. *Homeworkers in Global Perspective: Invisible No More*. New York: Routledge.

Brook, James, and Iain Boal, eds. 1995. *Resisting the Virtual Life: The Culture and Politics of Information*. San Francisco: City Lights.

Brunner, John. 1975. *The Shockwave Rider*. New York: Random.

Buchanan, Ruth, and Sarah Koch-Schulte. 2000. *Gender on the Line: Technology, Restructuring and the Reorganization of Work in the Call Centre Industry*. Policy Research Series. Ottawa: Status of Women Canada.

Bula, Frances. 2000. "Waterfront Plan Targets High Tech." *Vancouver Sun*, 17 October, B4.

Burawoy, Michael, and János Lukács. 1992. *The Radiant Past: Ideology and Reality in Hungary's Road to Capitalism*. Chicago: University of Chicago Press.

Butler, J., and J. Getzels. 1985. "Home Occupation Ordinances." Chicago: APA Planning Advisory Service.

Butler, Judith. 1993. *Bodies That Matter: On the Discursive Limits of "Sex."* New York: Routledge.

California. 1988. "California Statistical Abstract." Sacramento, CA: California Department of Finance.

Calthorpe, Peter. 1992. *The Next American Metropolis: Ecology, Community, and the American Dream*. Princeton, NJ: Princeton Architectural Press.

Canadian Telework Association. 1999. "Internet Penetration in Canada." <www.icv.ca>.

Carnoy, Martin, Manuel Castells, and Chris Benner. 1997. "Labour Markets and Employment Practices in the Age of Flexibility: A Case Study of Silicon Valley." *International Labour Review* 136 (1): 27-48.

Castells, Manuel. 1984. "Towards the Informational City? – High Technology, Economic Change, and Spatial Structure: Some Exploratory Hypotheses." Working paper #430, Institute of Urban and Regional Planning, University of California, Berkeley.

–. 1985. "High Technology, Economic Restructuring, and the Urban-Regional Process in the United States." In *High Technology, Space and Society*, edited by Manuel Castells. Beverly Hills, CA: Sage.

–. 1989. *The Informational City: Information Technology, Economic Restructuring, and the Urban-Regional Process*. Oxford, UK: B. Blackwell.

–. 1996. *The Information Age: Economy, Society and Culture*. Vol. 1, *The Rise of the Network Society*. Cambridge, MA: Blackwell Publishers.

–. 1997. *The Information Age: Economy, Society and Culture*. Vol. 2, *The Power of Identity*. Cambridge, MA: Blackwell Publishers.

Christensen, Kathleen. 1986. "Impacts of Computer-Mediated Home-Based Work on Women and Their Families." Draft. Washington, DC: Office of Technology Assessment.

–. 1988a. "Introduction: White-Collar Home-Based Work: The Changing U.S. Economy and Family." In *The New Era of Home-Based Work: Directions and Policies*, edited by K. Christensen, 1-11. Boulder, CO: Westview Press.

–, ed. 1988b. *The New Era of Home-Based Work: Directions and Policies*. Boulder, CO: Westview Press.

Chu, Clara. 2000. "The Digital Divide: A Resource List." UCLA Department of Information Studies, 17 September. <www.gseis.ucla.edu/faculty/chu/digdiv.htm>.

Chun, Miran. 1999. "There's No Place Like Home: Commuting by Computing." <www.infobeads.com>.

City of Vancouver Planning Department. 1996. *Live/Work and Work/Live: Vancouver Overview*. Vancouver, BC: City of Vancouver.

Cohen, Marjorie. 1991. *Women and Economic Structures: A Feminist Perspective on the Canadian Economy.* Ottawa: Canadian Centre for Policy Alternatives.

Conference Board of Canada. 1999. *Is Work-Life Balance Still an Issue for Canadians and Their Employers? You Bet It Is!* Ottawa: Conference Board of Canada.

Cooper Marcus, Clare. 1995. *House As a Mirror of Self: Exploring the Deeper Meaning of Home.* Berkeley, CA: Conari Press.

Corn, J., and B. Horrigan. 1984. *Yesterday's Tomorrows: Past Visions of the American Future.* New York: Summit Books.

Costello, Cynthia. 1988. "Clerical Home-Based Work: A Case Study of Work and Family." In *The New Era of Home-Based Work: Directions and Policies,* edited by K. Christensen, 135-45. Boulder, CO: Westview Press.

Cullen, J. et al. 1989. *Teleworking Application and Potential.* Dublin: NRC/NRB Dublin for E.C. Star Programme.

Dagg, Alexandra, and Judy Fudge. 1992. "Sewing Pains: Homeworkers in the Garment Trade." *Our Times* 11 (3): 22-5.

Davis, H. Craig. 1993. "Is the Metropolitan Economy Uncoupling from the Rest of the Province?" *BC Studies* (Summer): 3-19.

Davis, Mike. 1993. "Who Killed Los Angeles?" *New Left Review* 197 (January/February): 3-28.

Delaney, Paul. 1994. "Introduction: Vancouver as a Postmodern City." In *Vancouver: Representing the Postmodern City,* edited by P. Delaney. Vancouver: Arsenal Pul﹐ Press.

Deming, William G. 1994. "Work at Home: Data from the CPS." *Monthly Labor Review* 117: 14-20.

de Sola Pool, Ithiel, ed. 1977. *The Social Impact of the Telephone.* Cambridge, MA: MIT Press.

Despres, Carole. 1991. "The Meaning of Home: Literature Review and Directions for Future Research and Theoretical Development." *Journal of Architectural and Planning Research* 8 (2): 164-80.

DiMartino, Victor, and L. Wirth. 1990. "Telework: A New Way of Working and Living." *International Labour Review* 129 (6): 529-54.

Dovey, Kimberley. 1985a. "Home and Homelessness." In *Home Environments,* vol. 8, edited by I. Altman and C. Werner, 33-64. New York: Plenum Press.

–. 1985b. "The Quest for Authenticity and the Replication of Environmental Meaning." In *Dwelling, Place and Environment,* edited by David Seamon and R. Mugerauer. Dordrecht: Martinus Nijhoff Publisher.

Drucker, Peter F. 1988. "The Coming of the New Organization." *Harvard Business Review* (January/February): 45-53.

Duffy, Anne. 1997. "The Part-Time Solution: Toward Entrapment or Empowerment?" In *Good Jobs, Bad Jobs, No Jobs: The Transformations of Work in the 21st Century,* edited by Anne Duffy, Norene Pupay, and Daniel Glenday. Toronto: Harcourt Brace Canada.

Duffy, Anne, Norene Pupay, and Daniel Glenday, eds. 1997. *Good Jobs, Bad Jobs, No Jobs: The Transformations of Work in the 21st Century.* Toronto: Harcourt Brace Canada.

Duxbury, Linda, C. Higgins, and S. Mills. 1992. "Supplemental Work at Home and Work-Family Conflict: A Comparative Analysis." *Information Systems Research* 3: 173-89.

Edelstein, Michael. 1986. "Toxic Exposure and the Inversion of the Home." *Journal of Architectural and Planning Research* 3: 237-51.

Edwards, Paul, and Sarah Edwards. 1985. *Working from Home.* Los Angeles: Jeremy P. Tarcher.

Eichler, Margaret. 1995. *Change of Plans.* Toronto: Garamond Press.

Emmerson, Shannon. 1999. "'Coolness' Counts for Hip, Young Techies." *Vancouver Sun,* 29 July, E1.

European Foundation for the Improvement of Living and Working Conditions, Sheila Moorcroft, and Valerie Bennett. 1995. *European Guide to Teleworking: A Framework for Action.* Luxembourg: Office for Official Publications of the European Communities.

Faulkner, Wendy, and Erik Arnold, eds. 1985. *Smothered by Invention: Technology in Women's Lives.* London: Pluto Press.

Fischer, Claude. 1982. *To Dwell among Friends: Personal Networks in Town and City.* Chicago: University of Chicago Press.

–. 1985. "Studying Technology and Social Life." In *High Technology, Space and Society,* edited by M. Castells, 284-300. Beverly Hills, CA: Sage.

Fish, Susan, Kathleen Kurtin, and Catherine Nasmith. 1994. "Live/Work Opportunities in Garrison Common." Prepared for the City of Toronto Waterfront Regeneration Trust.

Forester, Tom. 1988. "The Myth of the Electronic Cottage." *Futures* (June): 227-40.

Foucault, Michel. 1980. *Power/ Knowledge: Selected Interviews and Other Writings 1972-1977.* Edited by Colin Gordon. Translated by Colin Gordon, Leo Marshall, John Mepham, and Kate Soper. New York: Pantheon.

Franck, Karen. 1989. "A Feminist Approach to Architecture: Acknowledging Women's Ways of Knowing." In *Architecture: A Place for Women,* edited by Ellen Perry Berkeley, 201-16. Washington, DC: Smithsonian Institution Press.

Frank, M. 1993. "Homework." *Journal of the American Planning Association* 59 (6).

Friedmann, John, and Goetz Wolff. 1982. *World City Formation.* Los Angeles: UCLA Comparative Urbanization Studies.

Fudge, Judy. 1991. *Labour Law's Little Sister: The Employment Standards Act and the Feminization of Labour.* Ottawa: Canadian Centre for Policy Alternatives.

Fuller, R. Buckminster, and Robert Marks. 1973. *The Dymaxion World of Buckminster Fuller.* Garden City, NY: Anchor Books.

Gannagé, Charlene. 1986. *Double Day Double Bind: Women Garment Workers.* Toronto: Women's Press.

–. 1995. "Restructuring and Retraining: The Canadian Garment Industry in Transition." In *Women Encounter Technology: Changing Patterns of Employment in the Third World,* edited by S. Mitter and S. Rowbotham, 127-49. London: Routledge.

Gerson, J., and R. Kraut. 1988. "Clerical Work at Home or in the Office: The Difference it Makes." In *The New Era of Home-Based Work: Directions and Policies,* edited by K. Christensen, 49-64. Boulder CO: Westview Press.

Giddens, Anthony. 1991. *Modernity and Self-Identity: Self and Society in the Late Modern Age.* Stanford: Stanford University.

Goffman, Erving. 1959. *The Presentation of Self in Everyday Life.* Garden City, NY: Doubleday Anchor.

–. 1963. *Behavior in Public Places: Notes on the Social Organization of Gatherings.* New York: Free Press.

Gordon, Gil, and M. Kelly. 1986. *Telecommuting: How to Make It Work for You.* New York: Prentice-Hall.

Gottlieb, Nina. 1988. "Women and Men Working at Home: Environmental Experiences." In *EDRA 19 Proceedings,* edited by Denice Lawrence. Washington, DC: EDRA.

Gough, Harrison, and A. Heilbrun. 1983. *The Adjective Checklist Manual: 1983 Edition.* Palo Alto, CA: Consulting Psychologists Press.

Graham, S., and S. Marvin. 1996. *Telecommunications and the City: Electronic Spaces, Urban Places.* London: Routledge.

Gram, Karen. 2000. "Coming Home." *Vancouver Sun,* 3 March.

Greater Vancouver Regional District. 1999. "Greater Vancouver Key Facts: A Statistical Profile of Greater Vancouver, Canada." Policy and Planning Department, Greater Vancouver Regional District, BC.

Greed, Clara. 1994. *Women and Planning: Creating Gendered Realities.* London: Routledge.

Gregory, Derek. 1982. *Regional Transformation and Industrial Revolution.* London: Macmillan.

Gringeri, C. 1996. "Making Cadillacs and Buicks for General Motors." In *Homeworkers in Global Perspective: Invisible No More,* edited by E. Boris and E. Prügl, 179-201. New York: Routledge.

Gurstein, Michael, ed. 2000. *Community Informatics: Enabling Communities with Information and Communications Technologies.* Hershey, PA: Idea Group Publishing.

Gurstein, Penny. 1990. "Working at Home in the Live-In Office: Computers, Space and the Social Life of Households." PhD diss., University of California, Berkeley.

–. 1991. "Working at Home and Living at Home: Emerging Scenarios." *Journal of Architectural and Planning Research* 8 (2): 164-80.

–. 1993. "Home-Based Businesses in the Greater Vancouver Regional District: Their Impact on Municipalities." Report. Greater Vancouver Regional District, Vancouver, BC.

–. 1995. *Planning for Telework and Home-Based Employment: A Canadian Survey on Integrating Work into Residential Environments.* Ottawa: Canada Mortgage and Housing Corporation.

–. 1996. "Planning for Telework and Home-based Employment: Reconsidering the Home/ Work Separation." *Journal of Planning Education and Research* 15(3): 212-24.

–. 1998. "Gender, Class, and Race in the Invisible Worksite: Disaggregating the Homeworker Population." *Gender, Technology and Development* 2 (2): 219-41.

Gutstein, Donald. 1999. *E.Con: How the Internet Undermines Democracy*. Toronto: Stoddart Publishing.

Habermas, Jurgen. 1986. "Modern and Postmodern Architecture." In *Critical Theory and Public Life*, 317-29. Edited by John Forrester. Cambridge, MA: MIT Press.

Hammer, Michael. 1990. "Reengineering Work: Don't Automate, Obliterate." *Harvard Business Review* (July/August): 104-12.

Haraway, Donna. 1990. "A Manifesto for Cyborgs: Science, Technology, and Socialist Feminism in the 1980s." In *Feminism/Postmodernism*, edited by Linda J. Nicholson. New York: Routledge.

Hareven, Tamara. 1977. "Family Time and Historical Time." *Daedalus* 106 (2): 57-70.

Harkness, R.C. 1977. *Technology Assessment of Telecommunications/Transportation Interactions*. Stanford, CA: Stanford Research Institute.

Harris Poll. 1999. "Time at Work and at Play." 10-15 June. <www.polling report.com/ workplay.htm>.

Harvey, David. 1989. *The Condition of Postmodernity: An Inquiry into the Origins of Cultural Change*. Oxford: Basil Blackwell.

Hatch, C. Richard. 1985. "Italy's Industrial Renaissance: Are American Cities Ready to Learn?" *Urban Land* (January): 50-9.

Hayden, Dolores. 1981. "What Would a Non-Sexist City Be Like?" In *Women and the American City*, edited by Caroline Stimpson, E. Dixler, M. Nelson, and K. Yatrakis. Chicago: University of Chicago Press.

–. 1984. *Redesigning the American Dream: The Future of Housing, Work and Family Life*. New York: W.W. Norton.

Hayward, Geoffrey. 1975. "Home as an Environmental and Psychological Concept." *Landscape* 20 (1): 2-9.

Heim, Michael. 1998. *Virtual Realism*. New York: Oxford University Press.

Helling, Amy. 2000. "Telework and the New Workplace of the 21st Century: A Framework for Understanding Telework." Washington, DC: US Department of Labor. <www.dol.gov/ dol/asp/public/telework/pl_3.htm>.

Herman, L. 1993. "Home Users Push PC Sales." *Electronic Business* 19: 63.

Hochschild, Arlie. 1989. *The Second Shift: Working Parents and the Revolution at Home*. New York: Viking Penguin.

–. 1997. *The Time Bind*. New York: Metropolitan Books.

Horwitz, Jamie. 1986. "Working at Home and Being at Home: The Interaction of Computers and the Social Life of Households." PhD diss., City University of New York.

Howard, Ebenezer. [1902] 1970. *Garden Cities of Tomorrow*. Cambridge, MA: MIT Press.

Hunter, Albert. 1974. *Symbolic Communities: The Persistence and Change of Chicago's Local Communities*. Chicago: University of Chicago Press.

Huws, Ursula. 1991. "Telework: Projections." *Futures* (January/February): 19-31.

–. 1993. *Telework in Britain*. London: Employment Department Research Series.

Huws, U., W. Korte, and S. Robinson. 1990. *Telework: Towards the Elusive Office*. Chichester, UK: John Wiley and Sons.

International Labour Organization (ILO). 1990. *Social Protection of Homeworkers: Documents of the Meeting of Experts on the Social Protection of Homeworkers*. Geneva: International Labour Organization.

–. 1995. *International Labour Conference 82nd Session Report V (1): Home Work – Fifth Item on the Agenda*. Geneva: International Labour Organization.

"Internet Users Spending Less Time with Families." 2000. *Vancouver Sun*, 17 February, A10.

Jacobs, Jane. 1961. *The Death and Life of Great American Cities*. New York: Random House.

Jameson, Fredric. 1984. "Postmodernism, or the Cultural Logic of Late Capitalism." *New Left Review* 146 (July/August): 53-92.

JALA Associates. 1985. *Telecommuting: A Pilot Project Plan*. Los Angeles: Department of General Services, State of California.

–. 1990. *The California Telecommuting Pilot Project Final Report.* Los Angeles: Department of General Services, State of California.

Japan Association for Planning Administration and Mainichi Newspapers. 1986. *International Concept Design Competition for an Advanced Information City – Competition Brief.* Kawasaki City: Japan Association for Planning Administration and Mainichi Newspapers.

Johnson, J. 1990. "21st Century Comes Home." *San Francisco Chronicle,* 18 March, F1.

Johnson, Laura. 1999. "Bringing Work Home: Developing a Model Residentially Based Telework Facility." *Canadian Journal of Urban Research* 8 (2): 119-42.

Joice, W. 1993. "Flexiplace: Getting the Job Done at Home (Government Employees)." *Public Manager* 22: 22-4.

Kawakami, S.S. 1983. *Electronic Homework: Problems and Prospects from a Human Resources Perspective.* Report LIR 494. Urbana-Champaign, IL: University of Illinois at Urbana-Champaign.

Koch-Schulte, Sarah. 2000. "Resistance of Tele-Service Workers: Implications for Qualitative Policy Research." Masters thesis, School of Community and Regional Planning, University of British Columbia, Vancouver.

Korte, W., S. Robinson, and W. Steinle, eds. 1988. *Telework: Present Situation and Future Development of a New Form of Work Organization.* Amsterdam: Elsevier Science Publishers.

Korte, Werner B., and Richard Wynne. 1996. *Telework: Penetration, Potential and Practice in Europe.* Amsterdam: IOS Press.

Krier, Rob. 1979. *Urban Space.* London: Academy Editions.

Lazarus, David. 1999. "Japan Doesn't 'Get' the Net." *Vancouver Sun,* 5 August, E1.

Leach, Belinda. 1993. "'Flexible' Work, Precarious Future: Some Lessons from the Canadian Clothing Industry." *Canadian Review of Sociology and Anthropology* 30 (1): 64-82.

Le Corbusier. 1929. *City of Tomorrow.* London: Architectural Press.

Lessinger, J. 1991. *Penturbia: Where Real Estate Will Boom after the Crash of Suburbia.* Seattle, WA: SocioEconomics.

Little, Jo. 1994. *Gender, Planning and the Policy Process.* London: Pergamon.

Loomis, John. 1995a. "Manufacturing Communities." *Places* 10 (1): 48-57.

–. 1995b. "Hotels Industriels." *Places* 10 (1): 24-6.

Lozano, Beverly. 1989. *The Invisible Work Force: Transforming American Business with Outside and Home-Based Workers.* New York: Free Press.

Lynch, Kevin. 1960. *The Image of the City.* Cambridge, MA: MIT Press.

McCamant, K., and C. Durrett. 1988. *CoHousing: A Contemporary Approach to Housing Ourselves.* Berkeley, CA: Habitat Press/Ten Speed Press.

McGregor, Douglas. 1985. *The Human Side of Enterprise: Twenty-fifth Anniversary Printing.* New York: McGraw Hill.

Mackenzie, Suzanne. 1988. "Building Women, Building Cities: Toward Gender Sensitive Theory in Environmental Disciplines." In *Life Spaces: Gender, Household, Employment,* edited by Carolyn Andrew and Beth Moore Milroy, 13-30. Vancouver: University of British Columbia Press.

Mahfood, P.E. 1992. *HomeWork: How to Hire, Manage and Monitor Employees Who Work at Home.* Chicago: Probus Publishing Company.

Markoff, John. 1990. "Virtual Death, Death and Virtual Funeral." *San Francisco Chronicle,* 9 September, F2.

Markusen, Ann. 1980. "City Spatial Structure, Women's Household Work, and National Urban Policy." *Signs: Journal of Women in Culture and Society* 5 (3, supplement): 23-44.

Mason, Roy, Lane Jennings, and Robert Evans. 1982. "The Computer Home: Will Tomorrow's Housing Come Alive?" *The Futurist* (February): 35-9, 42-3.

–. 1984. "A Day at Xanadu – Family Life in Tomorrow's Computerized Home." *The Futurist* (February): 30-5.

Menzies, Heather. 1981. *Women and the Chip: Case Studies of the Effects of Informatics on Employment in Canada.* Montreal: Institute for Research on Public Policy.

–. 1996. *Whose Brave New World?: The Information Highway and the New Economy.* Toronto: Between the Lines.

Meyrowitz, Joshua. 1985. *No Sense of Place: The Impact of Electronic Media on Social Behavior.* New York: Oxford University Press.

Micro-Credit Summit Draft Declaration. 1996. "Pre-Conference Document for the Micro-Credit Summit, February 2-4 1997." Washington, DC: Results Educational Fund.

Miles, Ian. 1988a. *Home Informatics: Information Technology and the Transformation of Everyday Life*. London: Pinter Publishers.

–. 1988b. "The Electronic Cottage: Myth or Near-Myth?" *Futures* (August): 355-66.

Miraftab, Faranak. 1996. "Space, Gender and Work: Home-Based Workers in Mexico." In *Homeworkers in Global Perspective: Invisible No More*, edited by E. Boris and E. Prügl, 63-80. New York: Routledge.

Mitchell, William. 1995. *City of Bits: Space, Place and the Infobahn*. Cambridge, MA: MIT Press.

Mitter, Swasti. 1986. *Common Fate, Common Bond: Women in the Global Economy*. London: Pluto Press.

–. 1992. *Computer-Aided Manufacturing and Women's Employment: The Clothing Industry in Four EC Countries*. London: Springer-Verlag.

–. 1998. "On Questioning of the Globality of Knowledge: Information Society and Women's World." In *Programme of Regional Conference – Gender and Technology in Asia*, 5-14. Bangkok, Thailand: Gender and Development Program, Asian Institute of Technology.

Mokhtarian, Patricia. 1991a. "Defining Telecommuting." *Transportation Research Record* 1305: 273-81.

–. 1991b. "Telecommuting and Travel: State of the Practice, State of the Art." *Transportation* 18 (4): 319-342.

–. 1995. "Country Report – USA." In *A Future of Telework: Towards a New Urban Planning Concept?* edited by F. van Reisen and M. Tacken, 97-8. Netherlands Geographical Studies 189. Utrecht: Delft University of Technology, Faculty of Architecture.

–. 1997. "The Transportation Impacts of Telecommuting: Recent Empirical Findings." In *Understanding Travel Behaviour in an Era of Change*, edited by P. Stopher and M. Lee-Gosselin, 91-106. Oxford: Elsevier.

Mokhtarian, Patricia, Narayan Balepur, Michelle Derr, Chaang-Iuan Ho, David Stanek, and Krishna Varma. 1997. "Residential Area-Based Offices Project: Final Report on the Evaluation of Impacts." Davis, CA: Institute of Transportation Studies, University of California, Davis.

Moore Milroy, Beth. 1991. "Feminist Critiques of Planning for Work: Considerations for Future Planning." *Plan Canada* 31 (6): 15-22.

Moore Milroy, Beth, and S. Wismer. 1994. "Communities, Work and Public/Private Sphere Models." *Gender, Place and Culture* 1 (1): 71-90.

Moss, Mitchell L. 1987. "Telecommunications, World Cities and Urban Policy." *Urban Studies* 24: 534-46.

Naisbitt, John. 1982. *Megatrends: Ten New Directions Transforming Our Lives*. New York: Warner Books.

Naisbitt, J., and P. Aburdene. 1990. *Megatrends 2000: Ten New Directions for the 1990's*. New York: Avon.

Nawodny, R. 1996. "Canadians Working at Home." *Canadian Social Trends* (Spring): 16-20.

Negroponte, Nicholas. 1995. *Being Digital*. New York: Knopf.

Nielsen Media Research and Canadian Broadcasting Corporation. 1998. "Internet Connections." <www.media-awareness.ca/eng/issues/stats/usenet.htm#Internet%20-Growth>.

Nilles, J. 1994. *Making Telecommuting Happen: A Guide for Telemanagers and Telecommuters*. New York: Van Nostrand Reinhold.

Nilles, Jack, with F. Roy Carlson Jr., Paul Gray, and Gerhard J. Hanneman. 1976. *The Telecommunications-Transportation Tradeoff: Options for Tomorrow*. New York: J. Wiley and Sons.

Norberg-Schultz, Christian. 1971. *Existence, Space and Architecture*. London: Studio Vista.

–. 1979. *Genius Loci: Towards a Phenomenology of Architecture*. New York: Rizzoli.

NUA Internet Surveys. 1999. "Internet Users by Location." <www.nua.net/surveys>.

Oakley, Ann. 1974. *Housewife*. London: Allen Lane.

O'Hara, Bruce. 1994. *Put Work in Its Place: How to Redesign Your Job to Fit Your Life*. Vancouver: New Star Books.

Olson, M.H. 1983. "Remote Office Work: Changing Work Patterns in Space and Time." *Communications of the ACM* 26 (3): 182-7.

Olson, M., and S. Primps. 1984. "Working at Home with Computers: Work and Nonwork Issues." *Journal of Social Issues* 40 (3): 97-112.

Ong, Aihwa. 1987. *Spirits of Resistance and Capitalist Disciplines: Factory Workers in Malaysia*. Albany, NY: SUNY Press.

Ontario District Council of the International Ladies' Garment Workers' Union and Intercede. 1993. *Meeting the Needs of Vulnerable Workers: Proposals for Improved Employment Legislation and Access to Collective Bargaining for Domestic Workers and Industrial Homeworkers*. Toronto: Intercede and Toronto Organization for Domestic Workers' Rights.

Orser, Barbara. 1993. "Hybrid Residential Development: An Exploratory Investigation of Multi-Use Residential Development." Ottawa: CMHC External Research Program.

Orser, Barbara, and M. Foster. 1992. "Home Enterprise: Canadians and Home-Based Work." Ottawa: Home-Based Business Project Committee.

Palm, R., and A. Pred. 1974. "A Time-Geographic Perspective on Problems of Inequality for Women." Working paper #236, Institute of Urban and Regional Development, University of California, Berkeley.

Park, Robert. 1952. *Human Communities*. Glencoe, IL: Free Press.

Pateman, Carole. 1989. "Feminist Critiques of the Public/ Private Dichotomy." In *The Disorder of Women*, edited by C. Pateman, 118-40. Cambridge: Polity Press.

Patton, Phil. 1993. "The Virtual Office Becomes Reality." *New York Times*, 28 October, C1.

Pearson, Ruth, and Swasti Mitter. 1993. "Employment and Working Conditions of Low-Skilled Information-Processing Workers in Less Developed Countries." *International Labour Review* 132 (1): 49-64.

Perkins Gilman, Charlotte. 1919. "But Here Is a House You Have Not Seen." *Ladies' Home Journal* 36 (February): 121.

Phizacklea, Annie, and Carol Wolkowitz. 1995. *Homeworking Women: Gender, Racism, and Class at Work*. London: Sage.

Postman, Neil. 1992. *Technopoly: The Surrender of Culture to Technology*. New York: Knopf.

Pratt, Joanne. 1993. "Myths and Realities of Working at Home: Characteristics of Home-based Business Owners and Telecommuters." Washington, DC: US Small Business Administration.

Pred, Allan. 1981. "Production, Family and Free-Time Projects: A Time-Geographic Perspective on the Individual and Societal Change in Nineteenth-Century U.S. Cities." *Journal of Historical Geography* 7 (1): 3-36.

–. 1984. "Place As Historically Contingent Process: Structuration and the Time-Geography of Becoming Places." *Annals of the Association of American Geographers* 74 (2): 279-97.

Prügl, Elisabeth. 1996. "Biases in Labor Law: A Critique from the Standpoint of Home-Based Workers." In *Homeworkers in Global Perspective: Invisible No More*, edited by E. Boris and E. Prügl, 203-17. New York: Routledge.

Prügl, Elisabeth, and Irene Tinker. 1997. "Microentrepreneurs and Homeworkers: Convergent Categories." *World Development* 25 (9): 1471-82.

Rabinow, Paul. 1982. "Space, Knowledge and Power: A Conversation with Michel Foucault." *Skyline* (March): 16-20.

–, ed. 1984. *The Foucault Reader*. New York: Pantheon Books.

Rapoport, Amos. 1969. *House, Form and Culture*. London: Prentice-Hall.

–. 1982. *The Meaning of the Built Environment: A Nonverbal Communication Approach*. Beverly Hills: Sage.

–. 1985. "Thinking about Home Environments: A Conceptual Framework." In *Home Environments*, edited by I. Altman and C. Werner. New York: Plenum Press.

Relph, Edward. 1976. *Place and Placelessness*. London: Pion

Rheingold, Howard. 1993. *The Virtual Community: Homesteading on the Electronic Frontier*. New York: Addison-Wesley.

Rifkin, Jeremy. 1995. *The End of Work: The Decline of the Global Labor Force and the Dawn of the Post-Market Era*. New York: G.P. Putnam and Sons.

Ritzdorf, Marsha. 1990. "Whose American Dream? The Euclid Legacy and Cultural Change." *Journal of American Planning Association* 56 (Summer): 386-90.

Robins, K., and M. Hepworth. 1988. "Electronic Spaces: New Technologies and the Future of Cities." *Futures* (April): 155-76.

Robinson, John. 1989. *The Rhythm of Everyday Life: How Soviet and American Citizens Use Time.* Boulder, CO: Westview Press.

"The Role of Self-Employment in U.S. and Canadian Job Growth." 1999. *Monthly Labor Review* (April).

Rosen, Laurel. 2001. "Where Have All the Lap Dancers Gone?" *Vancouver Sun,* 12 May, H8.

Rossi, Aldo. 1986. *The Architecture of the City.* Cambridge, MA: MIT Press.

Rowbotham, Sheila. 1993. *Homeworkers Worldwide.* London: Merlin Press.

Rowbotham, Sheila, and Swasti Mitter, eds. 1994. *Dignity and Daily Bread: New Forms of Economic Organizing Among Poor Women in the Third World and First.* London: Routledge.

Rowe, Colin, and Fred Koetter. 1985. *Collage City.* Cambridge, MA: MIT Press.

Saegert, Susan. 1980. "Masculine Cities and Feminine Suburbs: Polarized Ideas, Contradictory Realities." *Signs: Journal of Women in Culture and Society* 5 (3, supplement): S96-111.

Saegert, Susan, and Gary Winkel. 1980. "The Home: A Critical Problem for Changing Sex-Roles." In *New Space for Women,* edited by Gerda Wekerle, R. Peterson, and D. Morley. Boulder, CO: Westview Press.

Sale, Kirkpatrick. 1980. *Human Scale.* New York: Coward, McCann and Geoghegan.

Salomon, Ilan. 1985. "Telecommunications and Travel Substitution or Modified Mobility." *Journal of Transport Economics and Policy* 19: 219-35.

Sandercock, Leonie. 1998. *Towards Cosmopolis: Planning for Multicultural Cities.* New York: Wiley.

Sandercock, Leonie, and Ann Forsyth. 1992. "A Gender Agenda: New Directions for Planning Theory." *American Planning Association Journal* (Winter): 49-59.

Sassen, Sakia. 1991. *The Global City: New York, London, Tokyo.* Princeton: Princeton University Press.

Schneider, Jerry B., and Anita M. Francis. 1989. "An Assessment of the Potential of Telecommuting as a Work-Trip Reduction Strategy: An Annotated Bibliography." *Council of Planning Librarians (CPL) Bibliography* no. 246 (September).

Schumacher, E.F. 1973. *Small Is Beautiful: Economics As If People Mattered.* London: Blond and Briggs.

Scott, James. 1990. *Domination and the Arts of Resistance: Hidden Transcripts.* New Haven, CT: Yale University Press.

Scott, R. 1985. *Office at Home.* New York: Charles Scribner's Sons.

Seamon, David, and R. Mugerauer, eds. 1985. *Dwelling, Place and Environment: Towards a Phenomenology of Person and World.* Dordrecht: Martinus Nijhoff.

Sen, Amartya K. 1990. "Gender and Cooperative Conflicts." In *Persistent Inequalities,* edited by Irene Tinker, 123-49. New York: Oxford University Press.

Sennett, Richard. 1970. *The Uses of Disorder: Personal Identity and City Life.* New York: Random House.

Seron, C., and K. Ferris. 1995. "Negotiating Professionalism: The Gendered Social Capital of Flexible Time." *Work and Occupations* 22: 22-47.

Servon, Lisa. 1995. "The Economic Development Potential of Microcredit: Myths and Misconceptions." In *Local Economic Development in Europe and the Americas,* edited by C. Demaziere and P. Wilson. London: Mansell.

Shragge, Eric. 1992. "Community Based Practice: Political Alternatives or New State Forms." In *Bureaucracy and Community,* edited by L. Davies and E. Shragge. Montreal: Black Rose Books.

Slesin, Suzanne. 1986. "Plugged In! An Apartment for the Electronic Family." *New York Times,* 30 October, 19.

Smith, Ralph Lee. 1988. *Smart House: The Coming Revolution in Housing.* Columbia, MD: GP Publishing.

Stack, Carol B. 1997. "Beyond What Are Given As Givens: Ethnography and Critical Policy Studies." *Ethos* 25 (2): 191-207.

Statistics Canada. 1981. *Canadian 1981 Census.* Ottawa: Statistics Canada.

–. 1991a. *Canadian 1991 Census.* Ottawa: Statistics Canada.

–. 1991b. *The 1991 Survey of Work Arrangements.* Ottawa: Statistics Canada.

–. 1997. *Census of Canada 1996.* Ottawa: Statistics Canada.

–. 1999a. *Survey of Internet Use in Canada*. Ottawa: Statistics Canada.

–. 1999b. *Labour Market Changes in the 1990s*. Ottawa: Statistics Canada.

Stein, Maurice. 1962. *The Eclipse of Community*. New York: Harper and Row.

Sundstrom, E. 1986. *Work Places: The Psychology of the Physical Environment in Office and Factories*. New York: Cambridge University Press.

Swift, Jamie. 1995. *Wheel of Fortune: Work and Life in the Age of Falling Expectations*. Toronto: Between the Lines.

Tapscott, Don, and A. Caston. 1993. *Paradigm Shift: The New Promise of Information Technology*. New York: McGraw-Hill.

"Telecommuters in America." 1997. *USA Today*, 24 October, 9. Reprinted from *Detroit News*, <detnews.com/1997/cyberia/9710/23/10230112.htm>.

"Telecommuting: Staying Away in Droves." 1987. *Economist*, 4 April, 88.

Toffler, Alvin. 1980. *The Third Wave*. New York: William Morrow and Co.

–. 1990. *Powershift: Knowledge, Wealth, and Violence at the Edge of the 21st Century*. New York: Bantam.

Toffler, Alvin, and Heidi Toffler. 1995. *Creating a New Civilization: The Politics of the Third Wave*. Atlanta: Turner Pub.

Tonnies, Ferdinand. [1887] 1963. *Community and Society*. Trans. Charles P. Loomis. New York: Harper.

Totterhill, Peter. 1992. "The Role of Local Intervention: Choices and Agencies for Change." In *Computer-Aided Manufacturing and Women's Employment: The Clothing Industry in Four EC Countries*, edited by Swasti Mitter. London: Springer-Verlag.

Tuan, Yi-Fu. 1977. *Space and Place: The Perspective of Experience*. Minneapolis, MN: University of Minnesota.

Turkle, Sherry. 1995. *Life on the Screen: Identity in the Age of the Internet*. New York: Simon and Schuster.

Ulberg, Cy, et al. 1993. "Evaluation of the Puget Sound Telecommuting Demonstration: Survey Results and Qualitative Research." Olympia, WA: Washington State Energy Office.

US Department of Labor, Bureau of Labor Statistics. 1981. "Employment and Earnings." Washington, DC: Department of Labor.

US Department of Transportation. 1993. *Transportation Implications of Telecommuting*. Washington, DC: US Department of Transportation.

Waddell, Cynthia. 1999. "The Growing Digital Divide in Access for People with Disabilities: Overcoming Barriers to Participation in the Digital Economy." In *Understanding the Digital Economy: Data, Tools and Research*. Conference proceedings. US Department of Commerce, Washington, DC, 25-6 May. <http://icdri.org/the_digital_divide.htm>

Walter, Bob, L. Arkin, and R. Crenshaw. 1992. *Sustainable Cities: Concepts and Strategies for Eco-City Development*. Los Angeles, CA: Eco-Home Media.

Webber, Melvin M. 1964. "The Urban Place and the Nonplace Urban Realm." In *Explorations into Urban Structure*, edited by Melvin M. Webber, J.W. Dyckman, and D.L. Foley. Philadelphia: University of Pennsylvania Press.

Weisman, Leslie Kanes. 1992. *Discrimination by Design: A Feminist Critique of the Man-Made Environment*. Urbana, IL: University of Illinois Press.

Wexler, Mark. 1999. *Confronting Moral Worlds: Understanding Business Ethics*. New York: Prentice-Hall.

Wikström, Tomas, Karin Palm Lindén, and William Michelson. 1998. *Hub of Events or Splendid Isolation: The Home As a Context for Teleworking*. Stockholm: KFB.

Willey, M.M., and S.A. Rice. 1933. *Communication Agencies and Social Life*. New York: McGraw-Hill.

Williams, Frederick. 1982. *The Communications Revolution*. Beverly Hills: Sage.

Women and Work Research and Education Society and the International Ladies Garment Workers Union. 1993. *Industrial Homework and Employment Standards: A Community Approach to Visibility and Understanding: A Brief for Improved Employment Legislation for the Ministry of Women's Equality*. Vancouver, BC: Women and Work Research and Education Society and the International Ladies Garment Workers Union.

Wright, Frank Lloyd. 1932. *The Disappearing City*. New York: Payson.

Wright, Gwendolyn, and Paul Rabinow. 1982. "Spatialization of Power: A Discussion of the Work of Michel Foucault." *Skyline* (March): 14-15.

Zuboff, Shoshona. 1988. *In the Age of the Smart Machine: The Future of Work and Power.* New York: Basic Books.

Index